We Are One

THE SCIENCE OF ONENESS

The Sacred Science Chronicles:

Volume VI

ROBERT SIBLERUD

New Science Publications
9435 Olsen Court
Wellington, Colorado 80549
(970) 568-7323
www. Newsciencepublications.net

WE ARE ONE
The Science of Oneness

Published by:
New Science Publications
9435 Olsen Court
Wellington, Colorado 80549

Printed in USA
ISBN: 978-0-9666856-7-1
Library of Congress Control Number: 2011908895

Acknowledgments:

To My Good Friends

A True Visionary: **Maury Albertson, Ph.D.**
Foreword: **Robert T. Williams, Ed.D.**
Editing: **Margaret Shaw**
Editing: **Shirley Parrish**
Cover Art: **Ron Russell**
Layout: **Pat Alles**

Permissions

Dr. Tim Binder: The University of Science and Philosophy
Dr. Claude Swanson: *Life Force, The Scientific Basis*
Dr. David R. Hawkins: *Power vs. Force*

Other Books by Robert Siblerud

THE NEW SCIENCE CHRONICLES
Our Future is Hydrogen: Energy, Environment, and Economy – Volume I
Mind Your Health, Heal Your Body: A Guide to Wellness – Volume II

THE SACRED SCIENCE CHRONICLES
In the Beginning: Mysteries of Ancient Civilizations – Volume I
Keepers of the Secrets: Unveiling the Mystical Societies – Volume II
The Science of the Soul: Explaining the Spiritual Universe – Volume III
The Unknown Life of Jesus: Correcting the Church Myth – Volume IV
Our Galactic Visitors: The Extraterrestrial Influence – Volume VI

TABLE OF CONTENTS

TRIBUTE

to

Maury Albertson, Ph.D.

(August 30, 1918 – January 11, 2009)

This book is dedicated to the memory of Dr. Maury Albertson, a visionary who dedicated his life to raising consciousness about poverty, the environment, and New Science. I was fortunate to have Maury, a true mentor to me, as one of my best friends.

Maury was an engineering Professor Emeritus at Colorado State University. During his tenure, he became Director of Research at CSU and was responsible for attracting more research money to the university than any other university west of the Mississippi River.

While at CSU, Maury was in charge of designing what became the Peace Corps adopted by President Kennedy fifty years ago. Maury was nominated for the Nobel Peace Prize in 2008 for his work with the Peace Corps and Village Earth, an organization Maury founded to empower impoverished villages to self sufficiency.

Maury was co-founder and president of the International Association for New Science (IANS) that examined the science behind new paradigms such as the mind-body connection, alternative health care, the environment, spirituality, reincarnation, and metaphysics. Under IANS, Maury co-founded Hydrogen Now, an organization dedicated to raise awareness of hydrogen energy as the energy of the future.

Maury truly had an open mind and is missed by all who knew him.

Robert Siblerud

FOREWORD

by

Robert T. Williams, Ed.D

We live in a world of perception. We have the free will to give meaning to all things and all events. You can give any meaning to these words you are reading. You can give any meaning to the book you are holding. If you look up, you can give any meaning to what you see. If you hear something, you can give any meaning to what you hear. Any sensation you have, you give it meaning.

The *you* that gives any and all meaning to everything you experience is the physical body you took on for this incarnation. It is a life support system necessary to experience the effects of your agreement to separate from God or however *you* identify the Power greater than you.

The physical body itself is a projection of your mind's wishes. The screen for these projections is the world created by *your* mind and therefore seems real. *You* use a language of agreed upon definitions that are used to give proof of *your* choice for separation.

Our choices in this physical world are most often between separation and oneness. If there is anger, selfishness, and fear, the choice is of separation. If there is peace, unselfishness, and love, the choice is of oneness.

These truths are found in many spiritual teachings and practices, especially in a book that many readers of *We Are One* are probably unaware of, entitled *A Course in Miracles*, first published in 1975. Many concepts presented in *A Course in Miracles* are supported by scientific explanations laid out in *We Are One*. The source of information for the *Course* is Jesus, who gave it to Helen Schucman, psychology professor at Columbia University's College of Physicians and Surgeons in New York City.

It all began in the 1960s when Helen Schucman, with a

Jewish background, was given this information. Helen described herself in the Preface to the *Course* as a "psychologist, educator, conservative in theory, and atheist in belief who worked in a prestigious and highly academic setting." The head of Helen's psychology department announced that he was tired of the angry and aggressive feelings that he was encountering in life situations and said, "There must be another way." For some reason, as if on cue, Helen said she would help him find another way.

Shortly thereafter, Helen began to have highly symbolic dreams with many strange images. She also was introduced to a voice which made no sound but gave rapid inner dictation, which she took down in a shorthand notebook. One of the first things the voice said to her was, "This is a course in miracles." Helen emphasized it was not automatic writing. After writing down the information, Helen would read it to fellow psychology professor William Thetford, who would transcribe the dictation. The Voice told Helen that it was Jesus who was communicating with her. Helen was told that the *Course* was going to deal with universal spiritual themes saying, "A universal theology is impossible, but a universal experience is not only possible but necessary."

The *Course* identifies two realms: the realm of knowledge and the realm of perception. In the realm of knowledge, there are no thoughts that exist apart from God, because God and His creation share one Will. In the physical world, people are in the realm of perception. Helen was told that the realm of perception belongs to the world of time, of change, of beginning, and of ending. This realm depends on interpretation, not facts. In this realm, one finds birth and death, beliefs in scarcity, loss, and separation. What one sees and hears in the realm of perception appears to be real because it permits into awareness only what conforms to the wishes of the one who perceives. These perceptions create a world of illusion. Such a world needs constant defense because it is not real. This world of illusion is a dream world. God has provided humanity a way out of this illusion: the Holy Spirit which mediates between the two realms.

Helen was told, "The world we see merely reflects our own

internal frame of reference, the dominant ideas, wishes, and emotions in our minds." In other words we look inside first at our agreed upon definitions and see what kind of world we want to see and then project it to the outer world. We essentially are making up our reality as a reflection of our beliefs, especially our belief that we are separate from God. Misperceptions are used to justify our own mistakes. We often see a world full of evil and despair. What we see is not true. It is an illusion. As people recognize their perceptual error, they will learn to move beyond this misperception and practice forgiveness, the greatest learning aid to reverse misperceptions. The function of the Holy Spirit is to help humanity escape the world of illusion by teaching us to reverse our thinking and unlearn our misperceptions.

The *Course* defines sin as a lack of love, a separation from God. It also says Love (God) is all there is. Sin is a mistake to be corrected rather than an evil to be punished. Only at the level of the Christ Mind is Oneness possible, but *The Course* emphasizes that this Oneness has never been lost, can never be lost. The Self that God created, including you and me, needs nothing because it is "forever complete, safe, loved, and loving." It is the Holy Spirit that transforms misperceptions into perfect lessons in forgiveness that awaken us from our dream or illusion. **The miracle is the change of perception**.

The *Course* says the opposite of seeing through our physical eyes is the vision of Christ, which demonstates unity rather than separation, and love rather than fear. In this material world, forgiveness is a necessary correction for all the mistakes we have made. Through forgiveness, the thinking of the world is reversed. By acknowledging Christ in our brothers, we recognize the presence of Christ in ourselves. This is the principle of Oneness.

You will discover in *We Are One*, the science underlying *A Course in Miracles*. This book explains the science behind illusion. Anything that can be measured or any event that we experience is nothing but an illusion. Once humanity realizes we are all connected and we practice forgiveness, we are practicing unity consciousness. The *Course* says that forgiveness is understanding that all that seems to be consists of mutually agreed upon definitions and is really only an illusion. Therefore, it never really happened, so there is nothing to forgive in the first place.

As the Introduction to the *Course* says, A *Course in Miracles* can be summed up simply:

Nothing real [love] can be threatened.

Nothing unreal [illusion] exists.

Herein lies the peace of God.

Dr. Robert T. Williams is a Professor Emeritus of Education at Colorado State University and a facilitator for *A Course in Miracles.*

PREFACE

Some of the most complex mysteries of life can be explained simply. Simplicity is the key to understanding the world. If it is simple and clarifies, it is probably the truth. This also applies to science. As you will discover in this book, we are all connected and affected by everything else in the universe. There is no such thing as space or time. Any event or anything that can be measured is but an illusion. What connects us is empty space that has more energy in one cubic centimeter than all of the visible universe. As you will see, there have been many names given to this void, and from this void of empty space comes our material world.

If everything in the material world is an illusion, what is real? This book will attempt to show that the only real thing is our thinking mind. It is what creates our material world. Our thoughts create energy that is recorded in a vacuum of empty space, also called the quantum field or Mind Matrix. From this Mind Matrix we create our physical universe. The only thing in this universe is light, and quantum physics has shown that light is both wave and particle. Our thoughts are light energy that cause light waves in the Mind Matrix to form the physical world, like ripples in an ocean. Our thoughts create a ripple (physical object) which will eventually disappear back into the ocean (Mind Matrix). Visionary Walter Russell, as you will see, said that Light is God and God is Mind. God is creating through Mind. Man is also creating through the mind and Russell said that man is Mind, and man is God. This is a different view of reality from our everyday concept, but is supported by the science of quantum physics over the past century.

My life purpose has been to help raise consciousness of this New Science through conferences and writings. My books, both the *Sacred Science Chronicles* and *New Science Chronicles*, have documented some of the amazing research in New Science. In the late 1980s, along with Maury Albertson, Ph.D., former Director of Research at Colorado State University and co-founder of the Peace Corps, and Brian O'Leary, Ph.D., former Apollo astronaut and astrophysics professor at Princeton University and Cornell, we founded the International Association for New Science of which I am the director. For many years we convened New Science conferences as a vehicle to raise consciousness about the science of new paradigms. Many scientists have risked their professional reputations and careers to explore science beyond the confines of academia and traditional theories. It has been my philosophy that there is no separation between science and spirituality. As you will see in this book, both obey the laws of the universe. I dedicate this book to those pioneer scientists who have given us a new view of reality.

There is a field of Oneness that connects science, psychology, philosophy, and religion. Each discipline uses different terms to describe this Unity Consciousness. As you read this book, you will begin to understand that we live in a world of illusion, and how all objects and events are part of this illusion. From the field of Unity Consciousness arises this illusion of the material world. *We Are One* strongly suggests that nothing is real except our thinking mind that continually adds to this infinite sea of Oneness. As you read this book, you will find different terminologies to convey the idea of Oneness. Because there are different terms that explain the same concept, I have listed them below to help alleviate confusion. At this stage you may have your own idea of the listed term's definition, but by the end of the book you may also conclude that they are one and the same, and that we truly are One.

Terms of Oneness and their sources include:

Science

Empty Space – Layman's term
The Field – Scientific author Lynne McTaggart's definition
Vacuum – Traditional science

Ether – Early science
Implicate Order – David Bohm, quantum physicist
Morphogenetic Field – Rupert Sheldrake
Zero Point Field – Physics
Zero Point Energy – Physics
Torsion Field – Russian quantum physics
Quantum Field – Quantum physics
Subtle Energy – Holistic health
Life Force Energy – Holistic health
Light – Walter Russell, quantum physics

Psychology

Mind Substance – Walter Russell, first scientist to explain unity consciousness
Universal Mind – Walter Russell, Ernest Holmes
Sea of Knowledge – Walter Russell
Mind Matrix – Max Planck, quantum physics
Divine Matrix – Max Planck, popularized by author Gregg Braden
Divine Mind – Charles Fillmore
Collective Unconscious – Carl Jung, psychiatrist

Philosophy

Unity Consciousness – New Thought philosophy
Oneness – New Thought philosophy
Brahman – Hindu
Prana – Hindu
Dharmakaya – Buddhism
Suchness – Buddhism
Tao – Chinese
Chi – Chinese
Void – Eastern philosophy

A Course In Miracles

Atonement
Truth
Innocence

Peace
Non-separation
Wisdom

Religion
God
Love
Father
Source
The Absolute
All That Is
Unity

I hope you enjoy the book!

Robert Siblerud

PART ONE

The Science Underlying Consciousness

Chapter One

THE MIND MATRIX

All That Is

The purpose of this book is to provide evidence that we are all connected, that we are all One, and that we are all part of God. Spiritual teachings have long taught about the interconnection of everything, but science is now providing evidence of this Oneness. As we shall discover, there is no separation between science and spirit, as they are two aspects of the One. Humanity is on a path of evolution. Evolution is leading us to understanding the Law of Oneness, and once this happens, it is said, we will evolve from the third dimension to higher dimensions. We are living in exciting times because spiritual sources are saying that the 2012 window is the completion of a 26,000 year cycle that is bringing mankind to understand Oneness, enabling us to make the ascension to the fourth and fifth dimensions. The fifth dimension is the consciousness of Oneness. As we explore the science of Oneness, it is important to understand reality and how it relates to Oneness. Oneness begins in the mind and this is our connection to God. As you will discover, **mind is the whole universe.** The first eight chapters are based on scientific evidence behind the concept of Oneness and unity consciousness. The remaining chapters are based on revelations received by three spiritual pioneers of the 20th century – Ernest Holmes, Walter Russell, and Charles Fillmore, all who popularized the concept of oneness to the world and the science of the mind.

In 1944, Max Planck, the father of quantum theory, described a universal field of energy that connects everything in creation. He said, "All matter originates and exists only by virtue of a force . . . We must assume behind the force the existence of a conscious and intelligent mind. **The mind is the matrix of all matter**." The Mind Matrix bestows everything in our world and yet we are the creators of our world. The new physics provides the proof that our reality is created by the mind. It demonstrates that by looking at something as small as an electron, the observer can change its properties. Quantum physics shows the very act of observation is an act of creation; consciousness is doing the creating.

What is the Mind Matrix? Various names have been attributed to the matrix, with many not knowing the exact nature underlying the terminology. As we shall discuss, Walter Russell called it "mind substance." The Mind Matrix is empty space, the void, universal mind, and the vacuum. It is the ether that scientists have long been trying to discover. In Hinduism the void is called "Brahman," in Buddhism it is "Dharmakaya," while in Taoism it is the "Tao." In Chinese philosophy the word *chi* literally means ether, the energy that animates the cosmos and brings forth form. Indian philosophy refers to it as "prana." Physicist David Bohm calls it the "implicate order." Scientists today call it the "zero point field" and quantum physicists call it the "quantum field." Contemporary authors Gregg Braden and Lynne McTaggart name it the "Divine Matrix" and the "Field" respectively. There are many names for this matrix that is everything that is.

This quantum field is the source of our physical world and being, a continuous medium, that is present everywhere in space. Particles are local condensations of the field that form the material universe. These particles come and go and lose their individual character and dissolve back into the underlying quantum field. This is the illusion mechanism that will be discussed later.

The Zero Point Field

Both Gregg Braden (*The Divine Matrix*) and Lynne Mctaggart (*The Field*) have done masterful jobs in explaining the Mind Matrix or zero point field. Braden writes that since the early times of the ancient Greeks, there were those who believed in a universal field of energy that connects everything, and they called it *ether*. Both Aristotle and Pythagoras called ether the mysterious "fifth element of creation joining fire, air, water, and earth."

Sir Isaac Newton, the father of modern science, used the term *ether* to describe an invisible substance that permeates the entire universe, believing it was responsible for gravity. He also thought of it as living spirit. In the 19th century James Maxwell, who proposed the electromagnetic theory, described the field as a material substance of a more subtle kind than visible bodies. He said the field exists in those parts of space which seem to be apparently empty. The 1906 Nobel Laureate Hendrik Lorentz said, "I cannot but regard the ether, which can be the seat of an electromagnetic field with its energy and its vibration as endowed with a certain degree of substantiality, however different it may be from ordinary matter."

Einstein thought something would be discovered to explain what occupied the emptiness of space. He said that ether is necessary for the laws of physics to exist and that space without ether is unthinkable. Without ether, there is no possibility for standards of space and time. However, most scientists today wrongly believe that circumstances in our universe happen independently of each other. As we shall see, they do not.

During the latter part of the 20th century science began to find evidence of the ether. *Nature* published a scientific article in 1986 by E. W. Silvertooth, who reported a movement in the ether field. It was connected to the motion of Earth through space, just as had been predicted. This experiment and others began to give evidence that ether does exist, as Planck had hypothesized.

This energy field does connect everything in the universe, and it also is part of what it connects. As Walter Russell had written last century, the field is spread out through the universe, and when a ripple occurs in the field (sea of Mind Substance) physical objects will appear. Thinking or consciousness is the key to tap into this Mind Matrix of the universe. Quantum mechanics has shown the vacuum is not empty. There is no such thing as nothingness.

One of the chief architects of Quantum Theory was Werner

Heisenberg, who developed the Uncertainty Principle. The principle implies that no subatomic particles ever stay at complete rest, but are in constant motion. This activity is due to a ground state of field energy that is constantly interacting with all subatomic matter. This basic substructure of the universe is a quantum field that cannot be eliminated. The Uncertainty Principle refers to the uncertainty of measuring the physical properties of the subatomic world, a realm of probabilities. It is uncertain how subatomic particles will react, but physicists can give the probability how they will behave.

Quantum physics has found that all particles interact with one another by exchanging energy through other quantum particles. Particles appear out of nowhere and annihilate each other in less than an instant, 10^{-23} seconds. This creates a random fluctuation of energy without any apparent cause. These fleeting particles that are generated during the brief moment are called virtual particles, and they exist only during that exchange.

These subatomic interactions, when added across the universe, give rise to enormous energy. The energy is more than is contained in all the matter of the universe. This energy in the zero point field is called zero point energy. Its name is derived because fluctuations in the field are still detected in temperatures of absolute zero, the lowest possible energy state. Zero point energy is present in the emptiest state of space out of which no more energy can be removed. As a result of the Uncertainty Principle, there will always be residual jiggling due to virtual particle exchange. In other words, zero point energy is always present.

In 1911, Max Planck realized that empty space was full of activity. In the quantum world, quantum fields are not mediated by force, but through an exchange of energy that is redistributed in a dynamic pattern. Quantum mechanics has shown that we live in a universe that is a quantum sea of light always in motion. Because particles are always changing, the energy level can't be measured. The energy in the zero point field continues to act on the particles, so that they never come to rest, even at absolute zero. This zero point field is a matrix that connects all points in space by a force, such as gravity or electromagnetism. This force is represented by a ripple in the zero point field or waves. The ripple in the zero point field is considered that area in space where this charge and its effect can be detected.

A subatomic particle is nothing more than a little knot of energy (ripple) which briefly emerges and disappears into the underlying quantum field (sea). Quantum physics has shown that the particles (ripples) are transient and insubstantial. They cannot be separated from empty space. Einstein recognized that matter was nothing but a disturbance in the ether and that the only fundamental reality was the underlying field itself.

The emission and absorption of virtual particles occur with photons and electrons as well as other quantum particles in the universe. The zero point field is the repository of all fields, all energy states, and all virtual particles. After adding up all the variety of particles in the universe going in and out of existence, there is a vast inexhaustible energy lying in the background of empty space. Physicists estimate the total energy field in the zero point field to exceed all matter in the universe by a factor of 10 to the 40th.

The existence of the zero point field implies that all matter in the universe is interconnected by waves that are spread throughout space and time. This is the chi and prana of Eastern philosophy. Even the Old Testament refers to it in God's first dictum: "Let there be light," out of which matter was created.

Hal Putoff, Ph.D., was a laser researcher at Stanford Research Institute, at the time the second largest "think tank" in the world. Puthoff believed that if one could tap into the zero point field, one could have unlimited energy. Puthoff demonstrated that the stable state of matter depended on the existence of the dynamic interchange of subatomic particles with the sustaining zero point field. Scientists have shown mathematically that electrons lose or gain energy constantly from the zero point field. Electrons acquire this energy without slowing down by tapping into the fluctuations of empty space. The zero point field accounts for the stability of all matter. Puthoff showed that fluctuations of the zero point waves drive the motion of subatomic particles and that all the motion of all the particles of the universe in turn generate the zero point field. It is sort of a self-generating feedback loop across the universe. Puthoff says the zero point field is a self-regeneration grand ground state of the cosmos. It constantly refreshes itself and remains in a constant state, unless it is disturbed. He said this means that all the matter of the universe is literally connected to the farthest reaches of the universe through zero point waves.

The waves on the subatomic level are represented by period oscillations that move through the medium of the zero point field. The waves are encoded and are carriers of information. The interference patterns of the waves result in a constant accumulation of information and have a virtual infinitive capacity for storage. The subatomic waves of the zero point field are constantly imprinting a record of the shape of everything. The zero point field is a mirror image and record of everything that ever was. McTaggart described the vacuum as "the beginning and end of everything in the universe."

Scientists believe the zero point field can account for all the unusual characteristics of quantum mechanics. It explains the Uncertainty Principle, the duality of waves and particles, and fluctuations in the motion of particles. Puthoff and others began to wonder if the zero point field could also account for gravity. All matter at the subatomic level of quarks and electrons jiggle because of its interaction with the zero point field. A fluctuating charged particle will emit an electromagnetic radiation wave. The secondary field causes the attractive force, and Soviet physicist Andrei Sakharov believed it had something to do with gravity. Through mathematics, Puthoff found gravitational effects were entirely consistent with zero point particle motion, answering a lot of questions physicists otherwise could not answer, such as the weakening of gravity, why gravity can't be shielded, and why we have positive mass.

The zero point field can also account for the nonlocality of electrons, i.e., the appearance that they travel faster than the speed of light. Puthoff believed that if zero point fluctuation was the mechanism acting on subatomic particles that causes one particle to affect another, then every part of the universe would be in touch instantly.

Everything in the world is a collection of electron charges interacting with a background sea of electromagnetic and other energy fields. Einstein's equation of $E=MC^2$ was the equation for the amount of energy necessary to create the appearance of mass. There is only one energy that causes the appearance of material objects. There is no mass but only charge. Mass is energy. Puthoff and Bernie Haisch suggested that both inertia and gravitation are electromagnetic phenomena resulting from interaction with the zero point field. Haisch and colleagues published a scientific paper showing that the universe owes its very structure to the zero point field. They believed that the vacuum caused the particles to accelerate, which in turn caused them to agglutinate into concentrated energy, or what we call matter. Puthoff and Haisch proved one of the most fundamental laws of the universe, F=MA (Force = mass [inertia] times acceleration.

It is estimated that in one cubic centimeter of empty space, the amount of energy is greater than the total amount of energy of all the matter in the known universe. The zero point field has imprinted everything that has ever happened. It provides the reason for a stable universe. With this attribute, the zero point field may someday be tapped into for free energy and provide a source from which to extract information.

Physicist Fritjof Capra believes the distinction between matter and empty space finally had to be abandoned when it became evident that virtual particles came into being spontaneously out of the void and vanished into the void, without any nuclear or other strongly interacting particle being present. The discovery of the dynamic quality of the vacuum (zero point field) is seen by Capra and many physicists as one of the most important findings of modern physics. The zero point field exists always and everywhere. It is the carrier of all material phenomena. It is the void out of which the proton creates the pi-meson and other subatomic particles. The appearance and fading of particles are merely forms of motion of the field. All interactions in quantum field theory are pictured as the exchange of virtual particles. The stronger the force between the particles, the higher the probability of the exchange process.

The physical vacuum or zero point field contains the

potential for all forms of the particle world. These forms are not independent physical entities, but merely transient manifestations (ripples) of the underlying void (sea of energy). The vacuum of empty space is truly a living void, pulsating in endless rhythms of creation and destruction. The vacuum is far from empty, but contains an unlimited number of particles which come into being and vanish without end.

Quantum Physics 101

A quantum is defined as a "discrete quantity of electromagnetic energy." It is the "stuff" our world is made of. Quantum physicists discovered what we call the solid world is not solid after all. Quantum physics is the study of things on a small scale that explains the forces underlying our physical world. Newtonian physics is the physics of the observable universe explained by mechanical science when space and time are absolute. In 1900 Max Planck published his theory of the world as bursts of energy called *quanta*. His experiments showed that matter exists as probabilities and tendencies rather than absolutes. This suggested that our reality may not be what we think it is. Einstein discovered that time was relative rather than absolute. He said that time and space cannot be separated, but exist together as a fourth dimension. Space is what keeps everything from happening in the same place. Physicist Konrad Engle said if you took away the space between matter, everything in the universe would scrunch together into a volume no larger than a speck of dust. Einstein said, "space without ether is unthinkable." Planck discovered that everything is connected through a very real unconventional energy. He said that on the smallest level, atoms and subatomic particles, in fact, act as if they are connected.

In 1997, scientists split a photon into two separate particles with identical properties. The separated particles were placed in a specifically designed chamber and were transmitted for 14 miles. Each twin particle reached its destination, but on their journeys, each had made the same exact "choice" as its twin, acting as if they were connected. Physicists call this communication"quantum entanglement." Project leader Nicholas Gisin said that when twin photons are separated geographically, and if one of them is modified, the other photon automatically undergoes the same change. The connection between the photon was so complete that it appeared to be instantaneous. Gisin said the connection would be made instantly if it were a few meters apart or the distance of an entire universe apart. This instantaneous connection is called nonlocality.

There are two main characteristics in the Quantum Field Theory. The first is the fact that all interactions involve the creation and destruction of particles, such as the absorption and emission of photons. The second feature is there is a basic symmetry between particle and antiparticle, meaning that for every particle, there exists an antiparticle with equal mass and opposite charge. The antiparticle of the electron is called a positron. Pairs of electrons and positrons can be created spontaneously by photons and can be made to turn into protons in the reverse process.

One discovery made by quantum physicists is that if you break matter down into smaller and smaller pieces, such as electrons and protons, the traits of the object no longer exist. Physicists discovered that an electron can manifest as either a particle or wave. As a wave, it can do things a particle cannot do. If one electron is fired at a barrier with two slits, it can go through both slits simultaneously. When wavelike electrons collide with each other, they create interference patterns. The electron and all subatomic particles can manifest as either a particle or a wave. Light, gamma rays, radio waves, and x-rays can all change from wave to particle and back again. These subatomic substances that can change from particle to wave are called *quanta*. Physicists consider quanta the basic stuff from which the entire universe is made. **In other words, the universe is made of light.**

The amazing thing is that **the only time quanta ever manifest as particles is when we are looking at them.** When an electron is not being looked at, experiments suggest it always acts as a wave. Quanta only coalesce into particles when they are observed. **This strongly suggests that the mind is responsible for the entire material world.**

French physicist Alain Aspect discovered that some subatomic processes result in the creation of a pair of particles with identical or closely related properties. A positronium (an unstable atom) is composed of an electron and a positron (an electron with a positive charge). A positron is the electron's antiparticle, its opposite. The two eventually annihilate each other and decay into two quanta of light or photons. They travel in opposite directions and have the ability to shape shift from one kind of particle to another. Aspect found that no matter how far apart the photons travel, they will always be found to have identical angles of polarization. Polarization is the special orientation of the photon's wavelike quality as it travels away from its point of origin. Somehow the two photons can instantaneously communicate with each other so they know which angle of polarization to agree upon.

This faster than speed of light communication is known as nonlocality. At the subquantum level, where the quantum potential operates, locations cease to exist. All points in space become equal to all other points in space. It is meaningless to speak of anything as being separated from anything else. The nonlocal quality of the quantum potential explains the instantaneous connection between particles without violating Einstein's relativity ban on anything traveling faster than the speed of light. Because of the quantum potential, it permeates all of space, and all particles are nonlocally connected. Quantum physicist David Bohm claims that reality is where all things are part of an unbroken web and embedded in a space that was as real and rich in process as the matter that moved through it.

How can the waves of energy that transmit everything including television, radio, reflected light, and even thought, travel from one place to another? A medium must exist that conveys the vibration from one point to another. Something must occupy empty space. For us to be connected, something must be there to do the connecting. It is evident that there is a great web that connects everything in creation. There has to be something that the particles move through. The web that connects us is the zero point field or Mind Matrix, as Max Planck called it. Planck called the Mind Matrix the "matrix of everything." This matrix of energy reflects back to us, what we create in our mind, our emotions, thoughts, and beliefs. This field exists under three conditions, writes Gregg Braden:

1. Everything that exists is within the matrix. What we do influences other parts of the matrix.

2. The matrix is holographic, meaning any portion of the matrix contains everything else in the field. As we shall see, consciousness is holographic.

3. The third condition is that past, present, and future are instantly connected. The Mind Matrix is also the vessel of time. As Braden writes, if we understand these three principles, we become participants rather than victims of the forces that we do not understand and can not see.

Probability and Possibility

The wavelength quality of subatomic particles should more accurately be thought of as waves of probability. Quantum physicists discovered that one can never predict with certainty where a subatomic particle will be at a certain time. All that can be predicted is the probability. Most of the subatomic particles known today are unstable, meaning they disintegrate into other particles after a certain time. One can predict only the probability of decay after a certain time. Probability is the fundamental characteristic of atomic reality which governs all processes and the existence of matter. Subatomic particles do not exist with certainty at definite places, but show only tendencies to exist and atomic events show tendencies to occur.

All particles can move forward and backward in time, just as they can move left and right, up and down in space. They create fourth dimensional maps that are traced out in space. There is a complete symmetry with regard to direction and time. Particles can move forward in time or they can move backward in time. For every process, there is an equivalent process with the direction of time reversal and particle replacement by antiparticle.

The interaction of particles can be interpreted in terms of cause and effect only when the space-time diagrams are read in a definite direction. When they are interpreted as fourth dimensional patterns without defined direction of time attached to them, there is no "before" nor "after," therefore no causation.

The subatomic particles of matter have a dual quality by appearing sometimes as a particle and sometimes as a wave. It is the wave nature of electrons that accounts for the identity of atoms and their great mechanical stability. The dual quality of particles is exhibited by light which can take the form of electromagnetic waves or of particles. At the subatomic level, matter does not exist with certainty at "definite places" but shows only tendencies to exist. This is the same with atomic events that have only a tendency to occur. The tendencies are expressed as probabilities and are associated with mathematical quantities taking the form of waves.

This is why particles can be waves at the same time. They are probability waves. All the laws of atomic physics are expressed in terms of probabilities. At the quantum level, the solid material objects of classical physics dissolve into wavelength patterns of probabilities. The patterns represent the probability of interconnection and not the probability of things. As isolated entities, atomic particles have no meaning, but can be understood by interconnectedness. Quantum Theory thus reveals a basic oneness of the universe. Nature does not have isolated entities, but a complicated web of relations between various parts of the whole.

Both quantum physicist pioneers Niels Bohr and Werner Heisenberg believed the universe exists as an infinite number of overlapping possibilities. They described the quantum field as a kind of "quantum soup" with no precise location or state of being until something happens to lock one of the possibilities into place. It is the simple act of observation, a person's awareness, that locks it into place. This explains the behavior of quantum particles as they are observed in experiments.

Princeton physicist Hugh Everett III believes quantum physics theory could explain many worlds of interpenetrating parallel universes. The theory suggests that at any given moment of time, there are an infinite number of possibilities occurring, and that all possibilities already exist and are happening simultaneously. Each of these possibilities is happening in its own space and can not be seen by the other. These unique spaces are called alternative or parallel universes. Everett suggests that we already exist in each of these universes. In one universe a person might be suffering from a severe illness. In a change of consciousness and refocus, there may be a miraculous healing. In a parallel universe the world around him would look much like it did before, but the person would be healed. Braden believes we have the ability to purposely create the conditions that lock one possibility of our choosing into the reality of our lives. We do it through thoughts, feelings, emotions, and beliefs.

Particles are created and destroyed without end. They are in continual variation of energy potential and probability. Particle interactions give rise to stable structure which builds up the material world, but they do not remain static. Capra describes the whole universe as being engaged in endless motion or a continuous cosmic dance of energy. Photons, electrons, and protons are all stable particles until they are involved in a collision process and are annihilated. Each of the subatomic particles interacts incessantly by emitting and reabsorbing virtual particles. Quantum physics has demonstrated that movement and rhythm are essential properties of matter. All matter is involved in a continuous cosmic dance.

Author Michael Talbot concluded that without an observer, it must be accepted that the universe would not exist. He said that it must be assumed that the universe had built into it the potential for containing observers.

The Observer

In quantum physics, the universe is experienced as a dynamic, inseparable whole which always includes the observer in an essential way. The concepts of time and space, isolated objects, and cause and effect lose their traditional meaning from this point of view, writes Fritjof Capra.

Quantum physics has demonstrated that everyone plays a primary role in this participating universe, as we all are part of the equation. Consciousness creates, and the universe is the result of this consciousness. Planck said everything exists because of the intelligent mind which he called the matrix of all matter. We are part of the Mind Matrix. The act of focusing our consciousness is an act of creating. The process of having expectations is the precise force that creates something for us to see. All of us as a society create suffering or healing or peace or war. There are no other causes except our consciousness. We are part of everything because of our thoughts.

Quantum mechanics has shown that consciousness has a direct effect on all the subatomic particles of creation. Professor Emeritus John Wheeler, a colleague of Einstein at Princeton, believed that we might live in a world that actually was created by consciousness itself. Quantum theory states that **no elementary phenomenon is a phenomenon until it is observed.** Wheeler claims that the building blocks of the universe are acts of observer participation. The matrix is reality itself that contains all possibilities and the everlasting essence that connects all things.

Wheeler concluded from quantum physics that we do not view the universe as being constituted of matter and energy, but of pure information. He equates the "observer participation" in quantum experiments with "recognition" or meaning. Some scientists believe that consciousness should be thought of as pure information. It is a process that takes place within our minds and somehow causes a real and measurable effect in the physical world. When the human mind interacts with the reality of the photon, it appears that it is operating upon the "information picture" of the universe as a whole. Author Michael Talbot also believes the ultimate makeup of the universe seems to be pure information. He said information, not mass or energy, is the ultimate fabric of things. The universe processes the information that ultimately manifests as the photon and other subatomic particles in a highly organized manner. Dr. Edwin Fredkin from MIT believes, like Wheeler, that the ultimate substance of the universe is pure information. Fredkin goes a step farther. He believes that information is not purely a creation of the human mind but was there all the time, waiting for technology that would allow us to study it. The true path to understanding this phenomenon is through the mind of a conscious observer. Until his death, Einstein on the other hand was very disturbed by the notion that mind may have something to do with the ultimate reality of the world. He had difficulty imagining that the mind of a mouse could change the state of the world.

Wheeler and others in the scientific community have accepted the evidence that in the quantum universe, nothing exists until it is observed. Quantum theory suggests that intelligence extends all the way down to the quantum level of matter, as quantum particles process information in a unique way. There seems to be a sensitive web of communication taking place among the particles, causing some physicists to conclude that subatomic particles in their own way are conscious. The lesson of the new physics, writes Talbot, is that there is intelligence in all things and that we are part of the living fabric of the cosmos. The new physics has shown that the universe is interconnected and consciousness can be everywhere. We are part of time and space when we bring energy to conscious awareness through the act of perception, and we create new objects. In other words, by creating time and space we create our separation.

If consciousness operates at the quantum level, it would also reside outside of space and time. This means we, theoretically, would have access to information from the past and future. This also means that we could use the zero point field to influence events in the past, which later manifest in the present, such as the occurrence of a disease. There are many ways we could use our intention to change the probabilities in the future before they develop. Science has shown that intention can heal, as demonstrated by prayer research.

Princeton professor Robert Jahn and colleague Brenda Dunn also believe that subatomic particles do not possess a distinct reality until consciousness enters the picture. In fact, they believe that instead of discovering particles, physicists may actually be creating them with their mind. They cite a recently discovered particle called the *anomalon* whose properties vary from lab to lab. It is an anomalous reality depending on who finds it.

Capra states that in modern physics, the universe is experienced as a dynamic, inseparable whole, which always includes the observer. All natural phenomena are ultimately interconnected, and in order to explain any one of them we need to understand the other. Only approximations of nature are possible. He says that in modern physics, physicists have come to see that all their theories of natural phenomena, including the laws they describe, are creations of the human mind. Some scientists believe that the human race, through billions of acts of being both the observer and participant through the ages, have actually played a major role in creating both the universe and the laws of physics. Others conclude that the observers at all points in time are constantly self-organizing and guiding the developments of the universe through history. This is all possible in the Mind Matrix.

ChapterTwo

THE HOLOGRAPHIC UNIVERSE

Everything is Connected

The Mind Matrix, as we have discussed, is known by various names such as the zero point field, Mind Substance, Universal Mind, the void, ether, vacuum, the Divine Matrix, Brahman, Tao, just to name a few. This field contains the waves of everything in the material world past and present. Quantum physics has demonstrated that these waves are everywhere in the universe. One cubic centimeter of empty space contains more energy than all the material substance of the universe because of this sea of interpenetrating waves. It is our thoughts that create ripples in this sea of energy that in turn create the material world. All material objects are but illusions that will dissolve back into this energy of the zero point field. It appears that the only real thing is man's thinking mind that creates the material world.

David Bohm

Quantum physicist David Bohm was not satisfied with the loose ends that quantum physics did not answer, and he developed the theory of the holographic universe. His name for the zero point field was the "implicate order." Bohm was a colleague of Albert Einstein at Princeton University. Einstein was so impressed with Bohm that he predicted Bohm would be the one to discover the Unified Field Theory that would unite quantum physics with Newtonian physics. The holographic theory of the universe is probably not what Einstein had in mind.

Originally from Pennsylvania, Bohm attended Penn State College and received a Ph.D. at the University of California in 1943. Bohm was puzzled with legendary quantum physicist Niels Bohr, who lacked interest in interconnectedness that was demonstrated by quantum physics. Bohm's journey to explain interconnectedness began at the Berkeley Radiation Laboratory where he was researching plasmas. A plasma is a gas containing a high density of electrons and positive ions. Once these atoms were in a plasma state, they stopped behaving like individual atoms and started behaving as if they were part of a larger and interconnected whole. A vast number of electrons was able to produce effects that were surprisingly well-organized. The plasma constantly regenerated itself and even enclosed impurities similar to an immune response of a biological organism. In 1947 Bohm accepted a position at Princeton University and continued his research with plasma and the collective movement of electron "plasmons." Still bothering him was Niels Bohr's interpretation of quantum theory. In 1951 Bohm published a textbook on quantum theory that Einstein described as the best book for clearly presenting quantum physics.

Contradicting traditional thought, Bohm believed that particles such as electrons do exist in the absence of an observer. He proposed a new theory about a field he called "quantum potential." The influence of this field did not diminish with distance, as did gravity and electromagnetic fields. It was equally powerful everywhere. Bohm also believed that any effect could have an infinite number of causes. He reminded scientists to remember that no single cause and effect relationship was ever really separate from the universe as a whole.

His "quantum potential" theory indicated that the behavior of the parts was actually organized by the whole. Subatomic particles were not independent, but were part of a larger system, suggesting that wholeness was in some ways the more primary reality. This theory explained the behavior of electrons in plasma behaving like an interconnected whole. Bohm believed the whole system was undergoing a coordinated movement at the subquantum level at which the quantum potential operated. All points in space became equal to all other points in space. It was meaningless to speak of anything being separated from anything else. He termed this phenonmenon "nonlocality." Because this quantum potential permeated all of space, all particles were nonlocally interconnected, explaining the connection between twin subatomic particles acting in concert, no matter the distance. Bohm described a reality in which all things were part of an unbroken web embedded in space. Bohm's theory on wholeness and nonlocality received limited response at the time he proposed it.

Bohm discovered a device that could serve as a metaphor for his theory. This device involved a specially designed jar containing a large rotating cylinder. The narrow space between the jar and cylinder contained glycerine, a thick, clear liquid. A drop of ink was floated on the glycerine. When the handle of the cylinder was turned, the drop of ink was spread into the syrupy glycerine and seemed to disappear into the glycerine substance. However, if the handle was rotated in the opposite direction, the faint trace of ink began to reappear until it once again formed a droplet on the surface of the glycerine. Bohm saw that when the ink drop was spread out in the glycerine, its order was hidden in the glycerine. When it was on the surface of the glycerine, it had a different order. He called the hidden order the "enfolded order" and the manifest droplet the "unfolded order."

The Hologram

The next metaphor Bohm used to explain his theory and understand order was the hologram. Holograms were discovered in the late 1940s by Dennis Gabor who won a Nobel Prize for his efforts. The books, *The Holographic Universe* and *Beyond the Quantum,* written by Michael Talbot about David Bohm and the holographic universe has provided a great service to increase awareness of this remarkable pioneer-physicist. Much of this chapter is based on the writings of Michael Talbot.

A hologram is a three-dimensional photograph that is produced with the help of a laser. To create a hologram, one laser beam is shone on an object while a second laser beam is bounced off the reflected light of the first. The pattern of interference between the two beams is recreated on film to create a hologram.

When another laser is shown through the developed film, an image appears in all of its three dimensions.

If the hologram is cut in half and a laser beam is shone through one of the sections, it will continue to display the entire image of the original hologram. Every tiny section of a hologram contains its own bit of information, but also every other bit of information from the original holographic image. In a hologram, every part of the image interpenetrates every other part in the same way Bohm's nonlocal universe would penetrate all of its parts, writes Talbot. This concept has great implications. In a holographic universe, space and time are no longer viewed as fundamental. In other words, a hologram of the universe would contain past, present, and future, all enfolded and existing simultaneously.

Bohm believed that the universe was like a giant flowing hologram. The hologram allowed Bohm to see the universe as a cohesive whole, an idea which he published in the early 1970s. During the 1950s, Bohm had become a follower of the great Indian spiritual leader Krishnamurthi. Bohm believed that the hologram idea could also be applied to the spiritual realm. In 1980, Bohm published his book entitled *Wholeness and the Implicate Order.*

The Implicate and Explicate Orders

Bohm believed that the physical reality of everyday life was an illusion, similar to a holographic image. He thought there was a deeper order of existence, a primary level of reality that gave birth to all the objects and appearances of the physical world, similar to a piece of holographic film. Bohm called this deeper reality the implicate order or enfolded order. He referred to the physical world as the explicate order or unfolded order. The manifestations of all forms in the universe are the result of countless enfoldings and unfoldings between these two orders. Bohm believed that an electron was not one thing, but a totality enfolded throughout the whole universe. When an instrument detects a single electron, it is simply that one aspect of the electron has unfolded. The appearance of an electron is actually a continuous series of unfoldments and enfoldments. When a subatomic particle appears to be destroyed, it is not lost, but merely enfolded back into the deeper implicate order that we call the zero point field or Mind Matrix. Holographic film is like an implicate order because the image is encoded in its interference pattern in a hidden totality enfolded throughout the whole. The hologram that is projected from the whole represents the unfolded and perceptible version of the image.

The constant flowing exchange between the implicate and explicate order explains how particles, such as electrons in the positonium atom, can shape shift from one kind of particle to another. One electron enfolds back into the implicate order while another photon unfolds and takes its place. It also explains how a quanta can manifest as either a particle or wave.

The way an observer interacts with the field determines which aspect unfolds and which remains hidden in the implicate order. The term hologram implies a static quality, but Bohm thought of the universe as dynamic, describing it as a "holomovement." When something is holographic, no semblance of location exists. In the deeper implicate order, the holographic model explains why reality becomes nonlocal at the subatomic level. Saying that each part of a hologram contains all the information of the whole is another way of saying that the information is distributed nonlocally. Therefore, if the universe is a hologram, it would have nonlocal properties.

Bohm's theory claims that because everything in the cosmos is made of the seamless holographic fabric of the implicate order (Mind Matrix), it would be meaningless to view the universe as being composed of parts. An electron, for example, is just a name given to a certain aspect of the holomovement. Subatomic particles in the universe are not separate. Einstein said that time and space are not separate entities but are linked as a larger whole he called the space-time continuum. Bohm claims that everything is part of the continuum. Despite the apparent separation of things at the explicate order, each thing is but an extension of everything else. The implicate and explicate orders ultimately blend into each other. Bohm also asserts that everything is made of the same substance. Things can have their unique qualities and still be part of the divided whole. We go back to the metaphor of ripples and the ocean, where ripples (explicate order) will dissolve back into the ocean (implicate order). Bohm calls the different aspects of the holomovement "relatively independent subtotalities," not things. Bohm thinks dividing the world into fragments instead of seeing their interconnectedness, is at the root of many of today's problems.

Bohm also explains the effect consciousness has on the subatomic world through the holographic model. He believes that subatomic particles exist even when they are not observed, contrary to traditional quantum theory. He claims it has no meaning to speak of consciousness and matter interacting. The observer, he says, *is* the observed. The observer is the measuring device, the laboratory, the experimental results, and the wind blowing outside. Bohm believes that consciousness is a more subtle form of matter. (Walter Russell called matter "Mind Substance.") The relationship between matter and consciousness lies deep in the implicate order. Consciousness, Bohm says is present in various degrees of unfoldment and enfoldment in all matter. This explains why his plasma experiments showed traits of living matter, such as plasmons acting in concert similar to the immune system.

The ability of form to be active is the most characteristic feature of mind, claims Bohm. The electron is mindlike. Dividing the universe into the living and nonliving has no meaning. Bohm believes that even a rock is alive. Life and intelligence are present in all matter, space, time, and the entire fabric of the universe. Every portion of the universe enfolds the whole. "If we had knowledge how to access it, we could access the Andromeda Galaxy in the thumbnail of the left thumb," writes Talbot. Every cell in the body enfolds the entire universe, as does every cell in everything else. The whole past of the universe and implications of the future are also enfolded in each cell and small regions of space and time. We are One!

Bohm's implicate order is the foundation that has given birth to everything in our universe. It is his definition of the zero point field, the Mind Matrix, and the Tao. It contains every subatomic particle that has been or ever will be. It is the infinite ocean of energy where every region of space is awash in different kinds of fields, composed of waves of varying lengths. Each of these waves has energy. Talbot describes the energy as "a trillion atomic bombs in every cubic centimeter of life." This is zero point energy. Matter does not exist independently from this sea of energy that we call empty space. It is part of this space. Bohm believes we are also part of this fabric. We are a ripple in this ocean of energy. Bohm also conjectures that this implicate order is not the end of all things, but that there may be other undiscovered orders. Bohm is humble in believing that no theory is correct in an absolute sense, including his own. It is only an approximation of truth. He believes that future technology will be developed that will provide proof. Physicist Nobel Laureate Cambridge Briam believes Bohm's implicate order may someday lead to the inclusion of God.

The Holographic Brain

In the early 1940s, scientists believed that memories were stored in the brain. It was believed that memory had a specific location in the brain, somewhere in the brain cells. The mechanical model suggested that there was a one-to-one correspondence between certain types of memories and the location of the stored memories. The memory traces were called engrams, but no one knew what an engram was made of or where it was located.

Karl Pribram in 1946 worked for the renowned neuropsychologist Karl Lashley at Yerkes Laboratory of Primate Biology in Orange Park, Florida. Lashley was unable to produce any evidence of an engram. In his research, Lashley trained rats to perform a variety of tasks involving running through a maze. He then surgically removed parts of the brain and retested the rats. His goal was to eliminate the part of the brain that contained memory of running the maze. Lashley was dumbfounded when he found that no matter what part of the brain was removed, he could not eradicate their memories. Pribram believed the only answer was that memory was distributed throughout the brain as whole, but he could not account for the mechanism. In 1948, Pribram joined the faculty at Yale University.

Similar to the rat, human patients who had portions of their brain removed for medical reasons, never suffered the loss of specific memories. Even removal of parts of the temporal lobe did not create gaps in a person's memory.

Pribram read an article in the *Scientific American* in the mid-1960s about the hologram and then realized this might be the answer to memory. The key to the hologram is the phenomenon of interfering wavelengths, the crissscrossing patterns when two or more waves ripple through each other as in water. The brain is known to produce wavelengths such as alpha, beta, delta and theta waves. Perhaps this was the key to memory, Pribram thought.

In his rat research, Lashley had found the visual center of the brain was also resistant to surgical excision. Following removal of over 90 percent of the visual cortex, rats could still perform tasks that required visual skills. He found no one-to one correspondence between the image of the eye and its representation in the brain. Pribram concluded that the whole in any part as demonstrated by a hologram could explain this phenomenon.

His question was what wavelength phenomenon of the brain might be used to create the hologram. Pribram thought that because the neurons are packed so closely together, the expanding wavelength ripples of electricity were constantly crisscrossing one another. They were creating an endless array of interference patterns that might provide the brain its holographic properties and memories.

A holographic brain could explain so many things about memory, such as the storage of so many memories in so little space. The idea of a holographic brain could explain why some people have photographic memories. Pribram believed it could explain one's ability to transfer learned skills from one part of the body to another. It could explain the phantom limb experience, where some amputees still have sensations in the amputated limb.

Indiana University professor Paul Pietsch gave experimental evidence of the holographic brain. In a series of 700 operations on salamanders, he sliced, shuffled, and minced brains of salamanders. Whenever he replaced what was left of salamander brain back into the salamander, their behavior returned to normal. In 1979, Berkeley neurophysiologists Russell and Karen DeValois made a discovery that confirmed the holographic properties of the brain. They showed that the visual cortex was responding not to patterns but to the frequency of waveforms. Even after all this evidence of a holographic brain, the idea still remains controversial in the scientific community.

Pribram began to wonder, "If the picture of reality of our brain is not a picture at all but a hologram, what is it a hologram of?" He realized if the holographic model were taken to its logical conclusion, it would open the door to a different reality of the material world. Flowers, trees, leaves, and all other physical objects that we see would not exist in the way we believe they exist because holograms are virtual images, images that appear to be when they are not. A hologram seems to have extension in space, but one can pass a hand through it and feel no resistence. No instrument would pick up an abnormal energy where the hologram is located.

Pribram began to wonder if what the mystics had been saying for centuries was true, that reality was an illusion they called "maya." It appeared to Pribram that what was out there was really a vast, resonating sea of wave forms. It was a frequency domain that was transformed into the world as we know it after it entered our senses. Pribram then discovered the research and holographic theory of David Bohm and realized someone else shared his belief that we live in a holographic universe.

Chapter Three

IMPLICATIONS OF A HOLOGRAPHIC UNIVERSE

Our World As One

A holographic universe raises a multitude of questions. What is real? Is it possible that everything that we considered reality is but an illusion? Could the mystics be right, that our world is maya? Is the Mind Matrix the only real thing? Since thoughts create our reality, is our thinking mind the only real thing? What are the implications of being connection to everything? Would we change our ways if we understood the Law of One? These are very deep philosophical questions that the new physics has raised. Bohm believes that the universe is all thought and reality exists only in what we think. Our consciousness plays a role in the creation of subatomic particles. Bohm says there is no division between mind and matter because both are of the same substance. The brain is identical to mind and both are holographic. The implicate order implies a reality immensely beyond what we call matter, as matter is only a ripple in this Mind Matrix.

A Conversation With David Bohm

An interview with David Bohm was published in Ken Wilber's book entitled *The Holographic Paradigm* where Bohm explained the implications of a holographic universe. Bohm said the holographic model of consciousness is based on the premise that information from the consciousness is not stored in a particular place but stored throughout the brain. The more cells you bring together, the more detailed the information, meaning the more of the hologram you use, the more detailed and more ample the information will be. Any part of one has information about the whole. The selection of information is made by gathering together information from all over the brain similar to a hologram outside the brain. The hologram is an example of the enfolded or implicate order. Bohm said the whole enfolded order cannot be made manifest to us, but only some aspects can be made manifest. When we bring the enfolded order into the manifested aspect, we experience perception.

In the implicate order, said Bohm, the connection of the whole has nothing to do with locality in space and time. It has to do with an entirely different quality, chiefly unfoldment to the explicate order.

Bohm described the hologram as merely a fixed image of the state of the electromagnetic field. The holomovement is the ground of what is manifest. It is the movement of folding and unfolding. He explained that all existence is basically a holomovement which manifests in a relatively stable form. Everyone and everything are forms of holomovement.

The source of the generative matrix is the implicate order. He describes the brain function something like the implicate form manifested in consciousness by memory. Bohm said that consciousness is basically in the implicate order as is all matter. The implicate order is still matter. Consciousness manifests in the explicate order as matter. Consciousness, Bohm explains, is a more subtle form of matter or a more subtle aspect of the holomovement. Consciousness refers to thought, desires, and the whole mental or psychic aspect of life.

Thought is a material process because it is based on memory. It is the manifestation of deeper mind. Thought forms a world of its own in which it is everything. To imagine, Bohm claims, is also a form of thought. It is thought that has established a distinction between matter and spirit. Bohm describes spirit as that which is nonmanifest but moves through the manifest. *Holy*, Bohm said, merely means whole.

In the nonmanifest implicate order, all is one. There is no separation of space and time. Oneness also applies to ordinary matter in the explicate order, and even more so in subtle matter called consciousness. The whole purpose of the physical world is to experience separation so we can interact. In the nonmanifest reality, all is interpenetrating and connected in one. It can be said that the consciousness of mankind is one.

The basic problems of mankind are fear, jealousy, isolation, confusion, etc. Deep down, all the problems are the same. These problems originate in the consciousness of mankind and manifest in each individual. In other words, each individual manifests the consciousness of mankind, even though man perceives himself in the manifest as isolated, similar to the ripple in the ocean.

Thought is the source of disorder (ripple), not the content (ocean). All the problems lie in the nonmanifest, the pollution that has accumulated over the ages, called the sorrow of mankind. It is the nonmanifest consciousness of mankind. The universal sorrow consciousness creates an immense pressure to relieve it and this further corrupts and pollutes everything. It is through the individual that the general consciousness has to clear, said Bohm.

The individual is an actuality that manifests the consciousness of mankind. Everyone is in contact with the implicate order. Man is part of the whole of mankind, and in another sense he can get beyond it. Bohm said it takes more than a single higher energy of one saintly person to reach the whole of the consciousness of mankind. Bohm believes we need a more intense energy than a single individual can give. Bohm proposes that a number of individuals come together to help establish a one-mind consciousness of the whole. He said the effects of 10 or 100 people has a power immensely beyond one. This group consciousness can really ignite the whole consciousness of mankind. He said individual salvation has little effect on the whole. Nothing else can work better to increase consciousness than getting highly conscious people to act in concert.

Bohm believes that anything we put into thought is limited. The nonmanifest is much larger than the manifest but is still related to the manifest. The manifest and nonmanifest complete each other. He explained that love, compassion, intelligence, and insight are beyond both the manifest and nonmanifest. It is spirit or something unnameable.

Bohm said that consciousness is part of the whole. The whole of nature is in each part as well and consciousness is of that nature too. Consciousness is a material process. Consciousness is thought, feeling, desire, and will. However, Bohm said, consciousness is more subtle than the usual material process.

Man's reality is shaped entirely by ideas. Every idea is limited, and the idea can grasp some limited aspect of reality. Natural reality goes beyond any human idea, but the extent to which we can bring reality to our world depends on ideas. Knowledge held in the present is knowledge about the past, said Bohm.

The holomovement is an undefinable term, Bohm said. It has various factors and features such as light, electrons, sound, neutrons, etc., plus thought, feelings, desire, and will that we define as consciousness. Because they are interrelated, we can't reduce one of them to another.

The holomovement of each part is about the whole. No part contains all the details of the whole. It does not give a complete view of the past nor of the future. It only implies it.

We are all of the holomovement, and consciousness is a feature of the holomovement. The content of consciousness refers to the whole of the holomovement. If consciousness could empty itself of all its content (ripples), then possibly we could say the holomovement is then unobstructed. If we say consciousness is the manifest content, it is also the nonmanifest holomoverment underlying it.

Bohm said if consciousness empties itself of all these wrinkles (ripples), Krishnamurti would call it religion. It is the notion of ending conflict. Religion and wholeness means ending fragmentation and conflict or separateness.

The implicate order is where the holomovement takes place, an order that both enfolds and unfolds. Things are enfolded in the implicate order and the implicate order cannot be entirely expressed in an explicate fashion, said Bohm. People often speak of totality or wholeness which is both immanent and transient. In a religious context, this wholeness is given the name of God. Immanent means the totality of what is and transcendence means wholeness beyond matter.

Bohm said relativity and quantum mechanics have strongly

suggested that the world cannot be analyzed into separate and independent existing parts. Each part somehow involves all the others and enfolds them. This is one point on which all the mystics have agreed, i.e., the whole is present in each part at each level of existence. Bohm explained that mind is more subtle and is without matter. However, without matter, we have no knowledge of matter. It can be argued that mind is immanent matter, and in that sense matter is more comprehensive.

One of the weaknesses of some religions is that they have exalted the idea of spirituality and devalued matter. According to Bohm, we could say that all matter is in some way holy. One of the early teachings of mysticism stated, "As in spirit, so in matter," confirming the ancient hermetic view "as above, so below." This position can also be developed from modern physics.

According to quantum physics, said Bohm, everything is enfolded in everything else. As a result there is no separation of domains. Mind grows out of matter containing the essence of mind. Mind and matter are really both abstractions from the whole, said Bohm. They are relatively invariant subtotalities created by our thoughts.

Both mind and matter participate in the implicate order, said Bohm. The unfoldment of thought from the implicate is similar to the unfoldment of matter. We experience the implicate order both inwardly in our perception and outwardly in our physical movements.

Bohm asserts that if the paranormal exists, it can be understood through reference to the implicate order. There is no intrinsic reason why the paranormal should be impossible. The mystical experience goes still deeper into the implicate order, into the wholeness of mankind. It is the implicate order that provides commonality deep within matter, energy, light, and consciousness. The explicate order displays the implicate order. The implicate order would not be able to function without the explicate.

Bohm believes the implicate order is getting to know itself better. It is also reaching another level of consciousness, meaning another point in the evolution of consciousness. Quantum mechanics says there is an implicate order and an infinite sea of energy that unfolds to form, space, time, and matter. Bohm said matter, energy, and consciousness have an unknown common ground. Unlike the physicist, the mystic gives spiritual properties to this unity he experiences, calling it "meaningful, orderly, and good." The physicists would agree that it is orderly.

Thought reflects a reality other than itself, Bohm asserts. He believes we are fast approaching a state where thought reflects a reality that is itself. Mind has thought, feeling, desire, and will, but thought emphasizes the perception of categories, such as the universal or particular. Every phase of thought is naturally expressed through feeling. All these aspects flow into each other and implicate each other. When it comes to the thought of wholeness, says Bohm, thought will flow into the feeling of wholeness. Bohm defines wholeness as an undivided flowing movement, unbroken, and all-encompassing.

Spirit and matter are opposites which thought has produced based on perceived differences. They may be treated as independent, but actually, each is the principle movement of the other. Spirit becomes a principle movement of matter and vice versa. The condition of matter is the principle whereby spirit can act. If you look at the total movement, you can say matter and spirit are identical. They are one. Matter is saturated with spirit as spirit embeds itself within matter. They are completely interpenetrating. Bohm said, if you push matter to its furthest extent and understand its depth, you reach a cosmic consciousness and cosmic mind. It is cosmic order seen as an implicate consciousness.

Bohm said that quantum mechanics has shown a failure of the mechanistic order and the need for another order, the implicate order. This is parallel to the order we have observed in mind; therefore, it becomes possible to have a relationship between these two realms of mind and matter. Bohm said that if we can find a language in which mind and matter are seen as the same order, we might be able to discuss this experience intelligently.

By participating in the whole, human consciousness may give it meaning, said Bohm. We are an intrinsic feature of the universe, which would be incomplete without us in some fundamental sense. Only through our errors are we able to learn, to change ourselves, and to change everything. Bohm believed if we understand matter profoundly enough, it may turn out that experiences of the normal, paranormal, and the mystics may all have the same root in the implicate order.

We tend to see each object as separate and fixed, said Bohm, because we return to seeing the same object. The flowing movement regenerates the same thing over and over again causing us to lose sight of the movement itself. One of the basic features of matter is recurrence with great regularity and periodicity. All of our experiences, including thought, begin in immediate awareness of the flowing movement. If we experience thought and feeling, instead of naming and fixing them, feeling will flow into thought, and thought into feeling.

The relationship between two people is the moving principle of the other. We are linked to each other in the implicate domain and that is what we respond to. Mankind, Bohm said, is one implicate domain, both physically and mentally. The whole function of the explicate order is to display things as separate. The display becomes a guide to further activity. The activity is always one, but the display may be misleading. The other person becomes the moving principle of what we call "you," just as we are the moving principle of the other person. Matter may be the other person for us. We move it, and it moves us.

Bohm said the display in consciousness is a guide to the whole response of the body and the mind. Display means to unfold from the implicate to the explicate. If the display is confused, the results will be bad. If harmonious, the display will be orderly. With the power of thought, one can think oneself into illness or into health.

According to Bohm, evolution is a sign of creative intelligence of matter exploring different structure, which goes far beyond what is needed for survival. Evolution results from the creative movement of matter which is enfolded with intelligence. For example, in man, the mind plays a role in perpetuation of an aggravating illness, but with evolution it can also discover cures. So ends a conversation with David Bohm, who just may have discovered the elusive Unified Field Theory.

Evidence of a Holographic Universe

As we know, there are many unexplained phenomena in the material world that cannot be explained by Newtonian physics. By understanding quantum physics and holographic principles, we now have a good hypothesis underlying unexplained occurrences such as telepathy, the paranormal events, precognition, mysticism, oneness, and much more.

A hologram contains the original image in all of its many parts. A change made to just one of these portions becomes reflected everywhere thoughout the pattern. One mundane example is DNA. It is estimated that the average human has between 50 and 100 trillion cells in his or her body. Each cell holds 23 pairs of chromosomes that contain an individual life code of DNA. Each person has between 2,300 and 4,600 trillion copies of DNA in the body. When DNA does modify the blueprint of a species, it does not do it in a linear manner. Because of the holograph principle, when the DNA is modified, the change is reflected through the whole of the body. In this way, the subtle power of the hologram provides the leverage to make tremendous changes on a large scale by altering a pattern in just one place. This has great implications as we will discuss later.

Over six billion humans populate our planet. All have minds within the Mind Matrix or implicate order. Each individual mind is part of this large single consciousness. Every one of our minds contains the pattern of the whole consciousness and each also has direct access to this one consciousness. This gives us the potential to alter the hologram of the world. When like minds act in concert, the greater the effect and power like minds will have in changing this world.

In 1980, University of Connecticut professor Ken Ring, Ph.D., suggested that the near-death experience (NDE) could be explained by the holographic model. Ring was the president of the International Association for Near Death Studies as well as being the world's leading researcher of NDE. Ring believed that both death and NDE are nothing more than the shifting of a person's consciousness from one level of the hologram of reality to another.

Psychiatrist Stanislov Grof, an assistant professor of psychiatry at John Hopkins University School of Medicine in

1981 said only a holographic mind could explain such things as archetypal encounters with collective unconscious as well as other phenomena during altered states of consciousness.

Physicist Alain Aspect demonstrated that the web of subatomic particles that compose the physical universe shows holographic properties. His research demonstrated the nonlocality of twin particles that instantly mimicked what the other particle was doing, no matter the distance.

The holographic mind might help explain Carl Jung's concept of a collective unconscious. Jung believed that the dreams, artwork, fantasies, and hallucinations of his patients often contained symbols that could not be explained by their personal histories. The symbols closely resemble symbols of the world's great mythologies and religion. He believed these symbols sprang from the same source he called the collective unconscious. Jung termed these universal images "archetypes" that could be explained by the holographic model and the interconnectedness of all.

Bohm had written that "Deep down, the consciousness of mankind is one." However, in order to tap into this information of the implicate order, it has to directly be relevant to our memories. This is called personal resonance and as a result, only a few of the infinite images are available to us.

Montague Ullman, M.D., Professor Emeritus of clinical psychiatry at Albert Einstein College of Medicine believes that the ESP phenomenon of dreams can be explained by the holographic theory. He says dreams are a bridge between the perceptual and non-manifest orders that represent a "natural transformation of the implicate into the explicate." He also believes some aspect of psychosis can be explained by the holographic level of reality.

Schizophrenics often report oceanic feelings of oneness, but in a delusional way. They feel a loss of boundaries between themselves and others, believing their thoughts are no longer private. Many schizophrenics are able to read the thoughts of others. They often view people, objects, and concepts as members of larger subclasses, which seem to express holographic qualities of reality. Schizophrenics find sequential events meaningless, for all points in time are viewed as equal.

Ullman even believes aspects of holographic thinking can

explain symptoms of pronounced bipolar subjects. They can't keep up with all the thoughts and ideas that come at them in a manic state. Bipolar patients are often overwhelmed and have to manipulate those about them to accommodate their expanded vista. The result is confusion mixed with certain outbursts of creativity and success. As a result of the confusion, the manic becomes depressed.

Ullman also believes dreams are an aspect of the implicate order. Dreams, he believes, may provide a natural protection from coming into contact with more of the implicate order than they can handle.

Fred Alan Wolf, Ph.D., believes all dreams are internal holographs, and ordinary dreams are less vivid because they are virtual images. He believes the brain has the ability to generate real images, which it does during lucid dreaming. In a lucid dream, the dreamer is often able to control the dream in various ways, such as turning nightmares into pleasant experiences. In 1987, Wolf suggested that the holographic model could explain lucid dreams. Wolf believed lucid dreams are but actual visits to parallel realities.

Stanislov Grof, M.D., was a pioneer researcher of LSD, a hallucinogenic drug. He found that his LSD experimental subjects were able to unfold a deeper level of unconsciousness. Grof said he occasionally was able to cure schizophrenics by administering LSD. Patients on LSD were able to expand their consciousness beyond the usual boundary of the ego. They were able to explore what it was like to be other living things and even objects. For example, one of Grof's patients assumed the identity of a prehistoric reptile. Others were able to tap into the consciousness of their ancestral relatives. One patient was able to describe the ancient Egyptian practice of embalming and mummification. There seemed to be no limit to what Grof's LSD patients could tap into, including animals, disincarnate beings, spirit guides, etc. It was Grof who coined the term "transpersonal psychology."

Grof concluded that it was the holographic model that could explain the LSD results. It could explain the feeling that all boundaries are illusions; the lack of distinction between part and whole; and the interconnectedness of all things. These qualities would be expected in a holographic universe.

Another disorder that may be explained by the holographic universe is the multiple personality disorder (MPD), which shows earmarks of the implicate order. Victims of MPD have no awareness of their condition, with most averaging between eight and thirteen personalities. About 97% of MPD have suffered a severe childhood trauma, be it a psychological, physical, or sexual trauma. Each of the personalities possess a different brain wave pattern. Brain wave patterns are a global property of the brain suggesting a holographic process. A brain hologram, suggests Talbot, can store and call forth and assimilate a multitude of personalities. Perhaps what we call "self" is nothing but a hologram. The physiological changes that accompany MPD from personality to personality has profound implications. One personality may be diabetic while another is not. Another personality may be quite nearsighted while a second personality may have good vison. By changing a personality, one who is drunk may not be drunk with another personality. Very interesting!

Carl Jung defined synchronicity as "coincidences that are so unusual and so meaningful they could hardly be attributed to chance only." He discovered that synchronicities peak when the new realization or insight was just about to surface in a patient's consciousness. Jung was convinced that they were not chance occurrences but were related to the psychological process of the individual. Physicist Paul Davies believes the nonlocal effect is indeed a form of synchronicity because it establishs a connection between events. Physicist F. David Peat asserts that synchronicities are further evidence of the implicate order. He believes synchronicities are flaws in the fabric of reality that give us a glimpse of the large and uniting order underlying nature.

Bohm said that the apparent separation of consciousness and matter is an illusion, an artifact that occurs after both have unfolded into the explicate order of objects and sequential time. He emphasizes there is no division between mind and matter in the implicate order from where all things spring. Peat believes synchronicities reveal the absence of division between the physical world and inner psychological reality.

Dr. O. Carl Simonton, a radiation oncologist, developed a technique of mental imaging and visualization to help many cancer patients heal themselves. Simonton would have patients visualize white blood cells, the soldiers of the immune system, attacking and destroying cancer cells. Studies have shown this technique could extend life for cancer patients to twice the national norm. Psychologist Jeanne Achterberg, Director of Research and Rehabilitation at the University of Texas Health Service, believed the holographic theory of the brain could explain this phenomenon. Essentially, all experiences are ultimately just neurophysiological processes taking place in the brain, says Achterberg. The brain localizes and creates the internal hologram that we experience as reality. The brain is unable to distinguish between what we believe to be out there and what is really out there. If the brain acts holographically, the remembered image of a thing can have as much impact on the body senses as the thing itself, writes Talbot. The holographic model asserts that all experiences, whether imagined or real, are reduced to the same common language of holographic wave forms.

Bohm said every action starts from an intention in the implicate order. The imagination is the creator of the form. It already has the intention and germ of all the movement to carry it out. Creation takes place in the subtle level of the implicate order until it manifests in the explicate.

In the implicate order, as with the brain, imagination and reality are ultimately indistinguishable. Images in the brain can ultimately manifest as realities. Achterberg's research discovered that physiological effects produced through imaging are both powerful and specific. For example, Achterberg had her subjects imagine the increase of neutrophils, a specific white blood cell, and they did increase, while another group just increased T-cells. Achterberg believes that an ill person should recruit as many neural holograms of health as possible in the form of beliefs, well-being images, harmony, and specific immune functions. The healing process will accelerate. Conversely, she believes we should rid the mind of negative images because they can cause illness.

Another good example of the holographic universe is acupuncture. Acupuncture points in the ear, for example, are aligned in the shape of a miniature human. Dr. Terry Oleson, a psychologist at UCLA School of Medicine, believes there is a holographic nature to both the mind and body. He believes every portion of the body may contain the image of the whole, just like a hologram. Acupuncturists can work on the ear to affect the rest of the body. Dr. Ralph Alan Dale, director of the Acupuncture Education Center in North Miami Beach, Florida has evidence of 18 different micro-acupuncture holograms in the body including feet, hands, arms, neck, tongue, and even the gums. He even hypothesized that every finger and cell might contain its own acupuncture system. Iridology is another good example where the iris reflects the body's health in correspondence to anatomy. Palmistry and foot reflexology are other examples. Our body is truly a hologram.

Chapter Four

EASTERN PHILOSOPHY AND QUANTUM PHYSICS

Uniting Spirit with Science

Native Americans and other indigenous cultures around the world have talked about their oneness with nature and the Earth. Eastern mystics and Eastern religions have also taught that we all are one and connected. Even Western religions suggest there is a oneness to our world, emphasizing one God. These spiritual teachings seem to be saying the same thing that quantum physicists have discovered, that we are all connected. Bringing the greatest awareness to the similarities in the teachings of Eastern spiritual philosophy and quantum physics was Fritjof Capra, a professor of physics at the University of California, who wrote the classic book *The Tao Of Physics*. Many new thought scientists also believe there should be no separation between science and spirituality. Capra has used quantum physics to show the truths in these Eastern philosophies and religions. Much of this chapter is based on Capra's research in the interconnectedness of spirit and science. As you read this chapter, remember Brahman, Tao, and Dharmakaya are considered synonymous with the zero point field, Mind Matrix, void, Implicate Order, quantum field, etc.

Hinduism

For many centuries, Hindusim has influenced Indian intellectual life and shaped both the social and cultural life of the people of India. Capra said Hinduism can't be called a philosophy, nor can it be called a well-defined religion. He describes Hinduism as a large, complex socio-religious organism that consists of innumerable sects, cults, and philosophical systems. It involves various rituals, ceremonies, and spiritual disciplines. Hindus worship countless gods and goddesses.

The spiritual source of Hindusim is found in the Vedas, which is a collection of ancient scriptures written between 1500 and 500 BC by anonymous sages. Any philosophic system that does not accept the authority of the Vedas is considered to be unorthodox in India. There are four Vedas, the oldest being the Rig Veda written in Sanskrit, India's sacred language. Subsequent parts of the Vedas are concerned with sacrificial rituals that are connected with Vedic hymns, the last called the Upanishads. The Upanishads, Capra writes, contain the essence of the Hinduism spiritual message.

The teachings of Hinduism for the masses has been through popular epic tales, not the Upanishads. One epic, called the *Mahabharata,* contains India's favorite religious text, the *Bhagavad Gita,* a spiritual poem. *Gita* is a dialogue between the god Krishna and the warrior Arjuna. Krishna explains to Arjuna the different manifestations of the same ultimate reality. Krishna's spiritual teachings, as of all Hinduism, is the idea that the multitude of things and events around us are but different manifestations of the same ultimate reality, called Brahman. Brahman is the unifying concept that gives Hinduism its monastic quality, despite the worship of many gods and goddesses.

Brahman is considered the ultimate reality which is the "soul" or inner essence of all things. It is infinite and beyond all concepts. Brahman cannot be comprehended by the intellect nor adequately described. Some adjectives used to describe Brahman is beginningless, incomprehensible, the supreme soul, unlimited, unborn, unthinkable, not to be reasoned about, and beyond what is and beyond what is not. The sages of India consider Brahman divine and talk about it in mythological language. The various aspects of the divine Brahman have been given the name of various gods that are worshiped by the Hindus. However, scripture makes it clear that all these gods and goddesses are but reflections of the one ultimate reality. The manifestation of Brahman in the human soul is termed *atman*. The essence of the

Upanishads is the idea that *atman* and Brahman, the individual and ultimate reality, are one.

The Upanishads say, "That which is the finest essence - this whole world has that as its soul. That is reality. That is *atman*. That art thou." God becomes the world which, in the end, again becomes God. *Lila* is the name given to the creative activity of the Divine, the play of God. The world is seen as the stage of the divine play.

Brahman is the greatest magician who transforms himself into the world of matter. He performs this feat with his "magic creative power," which is the original meaning of maya in the Rig Veda. In Hinduism, maya has come to mean the pychological state of anybody under the spell of the magic play, *lila*. As long as we confuse the myriad forms of the divine *lila* with reality, without perceiving the unity of Brahman underlying all these forms, we are under the spell of maya.

Capra says maya does not mean that the world is an illusion, as is often stated. The illusion really lies in our point of view. The illusion is thinking that the shapes and structures, things and events around us are realities of nature, instead of realizing they are concepts of our measuring and categorizing minds. Maya is the illusion of taking these concepts for reality, confusing the map (ripple) with the territory (ocean), writes Capra. The Hindu's view of nature is that all forms are relative and ever changing maya, conjured up by the great magician (Implicate Order/Mind Matrix) of the divine play (Explicate Order).

The world of maya is always changing because the divine *lila* is a rhythmic, dynamic play. The dynamic force of the play is called karma, with karma meaning action. Karma is the active principle of the play. It is the total universe in action, where everything is dynamically connected with everything else. The *Gita* says, "Karma is the force of creation, where from all things have their life."

As long as we are under the spell of maya, writes Capra, and think we are separated from our environment and can act independently, we are bound by karma. **Being free from the bonds of karma means to realize the unity and harmony of all nature, including man, and to act accordingly.**

The *Gita* explains, "All actions take place in time by interweaving of the force of nature, but the man lost in selfish delusion thinks that he himself is the action. But the man who knows the relationship between the force of nature and action, sees how some forces of nature work upon other forces of nature, and becomes not their slave."

"To be free from the spell of maya, to break the bond of karma, means to realize that all the phenomena we perceive with our senses are part of the same reality, that everything including ourself is Brahman," writes Capra. The Hindu philosophy calls this experience *moksha* or liberation, which is the very essence of Hinduism.

There are many ways to achieve liberation. Hinduism does not expect all its followers to approach the Divine in a similar way. From this philosophy comes the great tolerance and inclusiveness which characterizes Hinduism.

The most intellectual school of Hinduism is called the *velanta* which emphasizes Brahman as a nonpersonal, metaphysical concept. This philosophy involves daily meditation and other spiritual practices to bring about the union with Brahman. Yoga means to "yoke or join" referring to the joining of the individual soul to Brahman.

For most Hindus, the most common path to approach the Divine is to worship the forms of personal gods or goddesses. There are thousands of deities. The three most worshiped gods include Shiva, Vishnu, and the Divine Mother.

Shiva represents the personification of the fullness of Brahman and can assume many forms. The god Shiva can impersonate many single aspects of the Divine. Shiva is the god of creation and destruction, who sustains through the endless rhythm of the universe (unfolding and enfolding from the implicate to explicate orders in scientific terms).

The god Vishnu appears under many guises including the god Krishna of the *Baghavad Gita*. In general, Vishnu's role is that of preserving the universe. The Divine Mother represents the many forms of the female energy of the universe. In Hindu religion/philosophy the body has always been considered to be an integral part of humans but not separated from the spirit.

In summary, the basic belief of Hinduism is that all these divinities are identical in substance. They are but manifestations of the same divine reality and reflect different aspects of the infinite, omnipresent, and incomprehensible Brahman.

Buddhism

The flavor of Buddhism is psychological, writes Capra, unlike the mythological and ritualistic flavor of Hinduism. Buddhism is considered exclusively with the human situation such as human suffering and frustration. Buddha's doctrine was not of metaphysics but of psychotherapy. Buddha identified the origin of human suffering and provided a solution to overcome it. Using the traditional concepts of maya, karma, and nirvana, Buddha provided a fresh, dynamic, and more relevant psychological approach. In India, Buddhism was absorbed and assimilated into the flexible Hindu religion. Buddha was adopted as an incarnation of Vishnu, the many-faced god.

There are two main schools of Buddhism:

1. *Hinayana* is the orthodox school closely following the teachings of Buddha. This sect is found primarily in Ceylon, Burma, and Thailand.
2. *Mayahana* is a more flexible school emphasizing the spirit of the doctrine which was considered more important. This sect spread to Nepal, Tibet, China, and Japan.

Buddha taught about the Four Noble Truths:

The First Noble Truth is called *duhkha*. It says the most outstanding characteristic of the human situation is suffering or frustration. The frustration results because humans have a difficult time facing the fact that everything around them is impermanent and in transition. Suffering, Buddha said, is when we resist the flow of life and become attached to fixed forms which are all maya, including things, events, people, and ideas. The doctrine of impermanence includes the belief there is no ego and no self, which is the persistent subject of our various experiences. The idea of ego or a separate individual self is but an illusion. It is another form of maya, which is an intellectual concept with no reality.

The Second Noble Truth concerns the causes of all suffering

called *trishna,* which is attachment or clinging. It is the futile grasping of life based on a wrong point of view called *avidya,* or ignorance. From this ignorance, we divide the world into individual and separate things. People attempt to confine the fluid forms into fixed categories created by the mind. As long as we separate things, we are bound to have frustration. If we cling to things that are transient, we will always be frustrated by not going with the flow. We are trapped in a vicious circle where every action generates further action. The vicious cycle is known as *samsara,* the rounds of birth and death, driven by karma. It is a never-ending chain of cause and effects.

The Third Noble Truth states that suffering and frustration can be ended. It is possible to transcend the *samsara* cycle and free oneself from the bondage of karma. Once a state of liberation is reached, we arrive in *nirvana.* Here, the oneness of all life has become a constant sensation. The false notion of a separate self has forever disappeared. *Nirvana* is a state of consciousness beyond all intellectual concepts. Once it is reached, it is called Buddhahood or awakeness.

The Fourth Noble Truth is the prescription to end all suffering. It is the eight-fold path of self-development which leads to Buddhahood. The path includes:

1. Right Understanding
2. Right Thoughts
3. Right Speech
4. Right Action
5. Right Livelihood
6. Right Effort
7. Right Mindfulness
8. Right Concentration

This doctrine is a means to reach enlightenment. Buddha emphasized the impermanence of all things, insisting on freedom from spiritual authority. It is up to the individual to achieve spiritual evolvement.

The *Mahayana* doctrine of Buddhism says that reality cannot be grasped with ideas or concepts. Buddha gave it the name *sunyata* or the void or emptiness. Once the futility of all conceptual thinking is recognized, reality is experienced as pure suchness. Reality or the void itself is not a state of mere nothingness, but is the source of all life and essence of all form.

Two elements compose the true enlightened wisdom by the *Mahayana* doctrine. Two pillars support this edifice of Buddhism:

1. *Prajna* which is transcendental wisdom or intuitive intelligence.
2. *Karuna* which is love or compassion.

The *Dharmakaya* is similar to Brahman. It pervades all material things in the universe and is also reflected in the human mind as *bodhi*, the enlightened wisdom. It is spiritual and material at the same time.

The emphasis on love and compassion are essential parts of wisdom. A *bodhi sutra* is a highly evolved human on the way to becoming a Buddha. He or she is not seeking enlightenment for oneself alone, but has vowed to help other beings achieve Buddhahood before entering *nirvana*. Buddha himself decided not to enter *nirvana* but returned to the world to show his path to salvation to his fellow mankind.

Buddhists have developed a doctrine of non-ego. Ego implies separation, and if there is no separate individual self, the idea of an individual entering *nirvana* does not make sense. This emphasizes the doctrine of oneness.

The culmination of Buddhist thought is the *Avaamsaka Sutra* which is the unity and interrelation of all things and events. This is the underlying philosophy of the Eastern world as well as the basic element of quantum physics. We are all interconnected.

Chinese Philosophy

The Chinese mystic unified himself in two complementary sides of human nature – intuitive wisdom and practical knowledge. During the 6th Century B.C., Chinese philosophy developed into two distinct schools, Confucianism and Taoism. Confucianism was the philosophy of social organization and practical knowledge. It helped establish a system of Chinese education and a strict protocol of etiquette. One of its main goals was to form an ethical basis for the traditional Chinese family system with its complex structure and rituals for ancestor worship. Taoism was primarily concerned with the observation of nature and the discovery of its way, called the Tao. Taoists believed human happiness was achieved when men followed the natural order by acting spontaneously and trusting their intuitive knowledge.

Confucianism was generally emphasized during the education of children who had to learn rules and convention that was necessary for life in society. Older people tended to pursue Taoism to regain their spontaneity that had been destroyed by social convention.

Like their counterparts in India, the Chinese believed there was an ultimate reality which underlay and unifed events and things of the material world. They believed reality was based on three principles – complete, all embracing, and the whole – referring to the one thing. They called this reality Tao, which originally meant "the way." It was the process of the universe, the order of nature. Tao was the ultimate, undefinable reality. It was equivalent to the Hindu Brahman and Buddhist Dharmakaya. The Tao was the cosmic process in which all things are involved. Taoists viewed the world as a continuous flow and change. By following the course of Tao, Chinese philosophies believed it was easy to manage the whole world by living in harmony with nature. The principle characteristic of the Tao was its cyclic nature and its ceaseless motion and change, of coming and going, of expansion and contraction.

Both Taoists and Confucianists believed in the doctrine of the Golden Mean. The Chinese believed that whenever a situation developed to its extreme, it was bound to turn around and become its opposite. This belief gave them courage and perseverance in distressed times and caution and modesty in times of success. Lao Tzu said, "The sage avoids excess, extravagance, and indulgence." The Chinese believed it was better to have too little than have too much. It was better to leave things undone than to overdo them.

The idea of cyclic patterns in the motion of Tao was given a definite structure by the introduction of the two polar opposites – yin and yang, writes Fritjof Capra. These are the two poles which set limits for the cycles of change. "The yang having reached its climax retreats in favor of the yin. The yin having reached its climax retreats in favor of the yang." All the manifestations of the Tao are generated by the dynamic interplay of these two polar forces. Yang represents the strong male-created power associated with heaven. Yin is the dark, receptive maternal element represented by Earth. Yang symbolizes movement and Yin is rest. The two dots in the Yin and Yang symbolize the idea that each time one of the forces reaches its extreme, it has within itself the seed of the opposite.

The Chinese believe that life is the blended harmony of yin and yang. The balance of yin and yang in the body and Earth is maintained by a continuous flow of chi or vital energy. Each organ has a meridian associated with it through which the chi energy flows. If the flow is blocked, the body falls ill. Acupuncture needles restore the flow of chi. The interplay of yin and yang are the principles that guide all the movement of the Tao.

The Chinese studied various combinations of yin and yang that developed into a system of cosmic archetypes. This system is represented in the I Ching, also called the Book of Change. The I Ching is the first among the six Confucian classics, and as a work, writes Capra, lies at the heart of Chinese thought and culture.

The starting point of the I Ching is a collection of 64 figures, or hexagrams, based on the yin and yang symbols. Each of the hexagrams consist of six lines, which may be broken (yin) or unbroken (yang). The 64 hexgrams constitute all possible combinations of yin and yang. These hexagrams are considered a cosmic archetype that represents the pattern of Tao in nature and in human situations. Each of the hexagrams represents an archetype category of the cosmos which is used in divination. The reason for consulting the I Ching is to predict the future and to help understand the present situation so proper action can be taken. The I Ching serves as both an oracle and book of wisdom.

Taoism is more interested in intuitive wisdom than rational knowledge, liberating one from the strict rules of convention. Taoists know that transformation and change are essential cornerstones of nature. All changes of nature are manifestations of the dynamic interplay of yin and yang. Taoists have developed two basic rules for human conduct:

1. When you want to achieve something, you should start with its opposite.
2. When you want to release something, you should admit in it something of its opposite.

Balance is important in Taoism. The Taoist sage does not strive for good, but tries to maintain a dynamic balance between good and bad. The Tao sage Lao Tzu found that all opposites are polar and therefore united. He said, "The way up and down is one and the same."

In Taoism, acting in harmony with nature means acting spontaneously and according to one's true nature. Taoists trust their intuition and intelligence. The Taoist adapts his or her action to the Tao's movement. By adopting the feminine, yielding qualities of human nature, it is easier to lead a more perfect, balanced life in harmony with Tao.

Zen

As with other Eastern religions and philosophies, Zen Buddhism of Japan also espouses the interconnection of nature and oneness. The pragmatic side of Chinese philosophy responded to Indian Buddhism by developing it into a spiritual discipline and gave it the name *chan,* meaning meditation. It was this *chan* philosophy that was adopted by Japan under the name of Zen. Essentially, Zen is the blend of the philosophies of three different cultures. It reflects the mysticism of India, the love of nature by the Taoists, and the pragmatism of Confusianism. The goal of Zen is the attainment of enlightenment. Zen philosophy believes that everyone has the potential to be awakened and to become enlightened.

Zen has no specific philosophy or doctrine and no formal creed. It encourages freedom from fixed beliefs, which makes it truly spiritual. Zen believes that words can never explain the ultimate truth. To perfect Zen is to live one's everyday life naturally and spontaneously. Great importance is placed on meditation for some Zen followers, many of whom will meditate for hours each day, while other Zen Buddhists believe enlightenment manifests itself in everyday affairs.

Eastern Philosophy and Unity

As we have seen, the most important essence of Eastern philosophy is the awareness of unity and the interconnectedness of all things and events. The Eastern view experiences all phenomena in the world as manifestation of a basic oneness. Eastern philosophies explain all things as interdependent and inseparable parts of the cosmic whole. They are but different manifestations of the same ultimate reality. It is called Brahman, Dharmakaya, and Tao. The Buddhists also refer to it as "suchness," writes Capra explaining that, "what is meant by the soul as suchness, is the oneness of the totality of all things, the great all includes whole."

In everyday life we are not aware of the oneness of all things, because we divide this world into separate objects and events. This is not a fundamental feature of reality, but a way of coping with life. The Hindus and Buddhists say this illusion is based on ignorance produced by the mind under the spell of maya.

Eastern mystics try to overcome this maya through meditation and quieting the mind. It is through meditation that the basic unity of the universe is experienced. Experiencing oneness is the fundamental characteristic of the mystical experience, and it is also one of the most important revelations of modern physics, writes Capra.

Quantum physics has shown that the constituents of matter and the phenomena involving matter are all interconnected, interrelated, and interdependent. They cannot be understood as isolated entities, but must be viewed as integral parts of the whole.

Explaining the science behind the Eastern view, Capra compares the mystic with the atomic physicist. Eastern sages say, "Things derive their being and nature by mutual dependencies and are nothing in themselves." Likewise, the quantum physicist says, "An elementary particle is not an independent existing unanalyzed entity. It is an essence, a set of relationships that reach outward to other things." Quantum theory shows us that the universe is not a collection of physical objects. It is a complicated web of relationships between the various parts of a unified whole. Likewise, the essence of the Buddhist *Avaamsaka Sutra* describes the world as a perfect network of mutual relations when all things and events interact with each other in an infinitely complicated way.

As in Eastern philosophy, quantum physics describes the universal interwovenness that always includes the human observer and his or her consciousness. At the quantum level, objects are understood only in terms of the interaction between the process of preparation and measurement during the experiment. At the end of this chain of process is always the consciousness of the observer. Measurements are but interactions, which create sensations in our consciousness. Heisenberg said, "What we observe is not nature itself, but nature exposed to our method of questioning." As we have seen, in quantum physics, the observer influences the properties of the observed object. Quantum theory has abolished the notion of fundamental separate objects, asserts Capra.

The Eastern mystic claims unity is reached in a state of consciousness where one's individuality dissolves into an undifferentiated oneness. This is where the world of the senses is transcended and the notion of things is left behind. Buddhists believe the external world and the internal world are only two sides of the same fabric. The thread of all forces, all events, all forms of consciousness, and all objects are woven into an inseparable net of endless, mutually conditioned relations, states Capra.

The Eastern sage understands the individuality of things, but also understands that the differences are relative within an all embracing unity. The unity of contrasts and opposites is one of the most interesting features of Eastern philosophy. Opposites are concepts that belong to the realm of thought and therefore are relative. By focusing on any one concept, we automatically create the opposite. The Chinese philosopher Lao Tsu said, "When all the world understands beauty to be beautiful, then ugliness exists; when all understand goodness to be good, then evil exists." The Eastern sage understands that good and bad, life and death, pleasure and pain are not absolute experiences belonging to different categories but are only two sides of the same reality. They are but extreme parts of a simple whole. The awareness that all opposites are polar and thus a unity is one of the highest goals of followers of Eastern tradition. The basic principle of Eastern philosophy is to view polar opposites as merely different aspects of the same phenomenon. A virtuous person in the East is one who is able to maintain a dynamic balance between good and bad. The unity lying between yin and yang is the Tao which is a process that brings about their interplay.

In Eastern mysticism, a unity between the feminine and masculine aspects of human nature is sought. Lao Tzu said, "A fully realized human is one who knows the masculine and yet keeps to the feminine." The main purpose of meditation in Eastern tradition is to balance the masculine and feminine mode of consciousness.

The unification of opposites in quantum physics is seen when subatomic particles are both destructible and indestructible, when quanta are both wave and particle, when matter is both continuous and discontinuous, and force and matter are but different aspects of the same phenomenon. All these opposites are transient in modern physics. Likewise, the Eastern sage says the complementary of opposites by the yin and yang and their dynamic interplay is the essence of all natural phenomena and all human situations.

Capra said that perhaps the Tao, Brahman, and Dharmakaya can be seen as the ultimate unified field from which spring all phenomena. This is the zero point field, the Implicate Order, the quantum field, and the void. The Upanishads say, "Brahman is life. Brahman is joy. Brahman is the void . . .The void, verily, that is the same as joy . . ." Buddhists call the ultimate reality *sunyata,* which means emptiness or the void. It affirms that "It is a living void which gives birth to all form in the phenomenal world." Likewise, the Taoist ascribes a similar infinite and endless creativity to the Tao and again calls it empty, describing heaven as empty and formless. The Tao has the potential of containing an infinity of things. The Eastern void has an infinite creative potential similar to the quantum field.

The phenomenal manifestation of the mystical void is dynamic. It comes into being and then vanishes in one ceaseless dance of movement and energy. There is a continuing birth and death. This dynamic describes David Bohm's holomovement of the implicate and explicate order. In Buddhist philosophy, the existence of any material substance and the idea of constant self undergoing successive experience is an illusion because they are part of the cosmic dance.

In Chinese philosophy, chi was the term used to describe the life force animating the cosmos. Chi is considered as a tenuous and nonperceptible form of matter which is present throughout space. It can condense into solid material. Philosopher Chang Tsai said, "When the chi condenses, its visibility becomes apparent, so there are these shapes (of individual things). When it disappears, its visibility is no longer apparent and there is no shape. The great void cannot but consist of chi; this chi cannot but condense to form all things; and these things cannot but become dispersed so as to form once more the great void." Capra writes, "In quantum field theory, the field (or chi) is not only the underlying essence of all material objects, but also carries their mutual interaction in the form of waves."

The Dance of Shiva is the symbol of the cosmic cycle of creation and destruction and the rhythm of birth and death. Shiva's dance is the dance of subatomic matter.

The Eastern mystic continuously has said that all things and events are creations of the mind that arise from a particular state of consciousness. It is our thinking that creates form out of the void, the Mind Matrix. As we have seen, quantum physics supports the ancient teachings of Oneness of Eastern philosophy.

Chapter Five

EVIDENCE OF THE MIND MATRIX

Power of the Mind

The theories espoused so far about the power of the mind seem to make sense, but is there any hard science to substantiate it? If the mind is creating our reality, it is the most powerful energy in the universe. As we have seen, some scientists believe it is the only energy that has created our world. Thank goodness there have been pioneer scientists who have researched the Mind Matrix to show the great potential of the mind. The ultimate reward of unified mind power would be to create a world of peace and harmony. As we shall see, there seems to be an exponential effect of like minds focusing on the same thing. Religions teach the power of prayer, and many studies have confirmed its positive effect on health. Some believe thoughts are prayers. If we could become the master of our minds and not slaves to wrong thinking, what a wonderful world we could create!

Random Event Generator

In 1979 at Princeton University, the Princeton Engineering Anomalies Research (PEAR) lab was founded by Robert Jahn, who was a professor of aerospace sciences and Dean Emeritus of the School of Engineering and Applied Science. Brenda Dunn, a clinical psychologist, was the lab director. After nearly a decade of rigorous experimentation they accumulated undeniable evidence that the mind can psychically interact with physical reality. Their research showed that through mental concentration alone, humans are able to affect the way certain kinds of machines operate. They were able to explain this by the holographic model. Their research provided compelling evidence for the existence of psychokinesis.

Jahn and Dunne employed a device called a random event generator referred to as an REG. An REG is able to produce a series of random binary numbers. Some describe it as an automatic coin flipper capable of producing an enormous number of coin flips in a very short time. The odds are that if a coin is flipped 10,000 times, the result would be an even split between heads and tails. The research protocol involved volunteers sitting in front of the REG and concentrating on it to tip the balance of either heads or tails. After hundreds of thousands of trials, they showed that through concentration alone, the subjects did have a small but statistically significant effect on the REG output. This ability to produce psychokinesis output was present in the majority of tested volunteers. Different subjects produced different but consistently significant results. In a similar protocol using a pinball machine, subjects were able to influence the balls to fall into a certain bin.

Because all known physical processes possess a wave/particle duality, Jahn and Dunne thought it reasonable to assume consciousness does as well. When consciousness was particle-like, it would be located in our head. Jahn and Dunne believed that only the reality created by the interaction of multiple consciousnesses is real. They believed that reality is established only in the interaction of consciousnesses with the environment.

Dean Radin, Ph.D., of the Heartmath Institute in California conducted research with the REG during the O.J. Simpson trial. There were three REG devices involved in the protocol, one at PEAR, one at the University of Nevada, and one at the University of Amsterdam. As many remember, there were millions of people glued to televison sets as the O. J. verdict was rendered. At the moment the verdict was read, the REGs recorded significantly higher peaks. Like everyone in the world, the REG computer was also brought to attention to find out whether O. J. was innocent or guilty. This suggested to Radin there may be a collective consciousness.

Radin and his colleagues had observed in their research the

apparent ability of the human mind to extend its boundaries. They began to wonder whether the effects of thought become larger when many individuals operate in unison and whether a collective mind could operate as a unity. If coherence could develop between individuals and their environment, would it be possible to produce a group coherence?

It was Roger Nelson, a college professor in psychology, who first designed an experiment to determine if an REG machine could provide evidence of a collective conscious. While at Princeton, Nelson became involved with PEAR and the REG machine in the civil engineering department. His research showed that when a group of people focused their attention, they could affect the REG device. This confirmed the premise, "We think, therefore we affect."

Nelson next researched his hypothesis with a device call "cont - REG," an abbreviation for keeping an REG running continuously to see if it registered any more heads or tails than usual during the ordinary course of a day and establish what was happening in the room the moment of effect registered by the REG.

Nelson discovered that a bonded couple who were intensely involved profoundly effects the REG device. This suggested that two like-minded people created more order in a random system.

His next question was what would happen if you assembled an entire crowd to focus on the same thing? He wanted to know if there was a correlation between crowd size and intensity of interest on the REG effect. He wanted to determine if there was a "conscious field" and he termed his research "field REG."

Nelson began his research on smaller events such as business meetings, academic meetings, concerts, theaters, etc. Nelson discovered the field REG was moving out of its random movement during moments of peak attention. The effects were relatively large, about three times greater than what individuals at PEAR were doing on their own. He found the activity did not matter, but it was the intensity of the group, especially if there was a group resonance. The most important criterion was whether their attention was focused on the same thought.

What appeared to be happening was when attention focused

the waves of individual minds on something similar, there was a type of super-group radiance occurring at the quantum level that produced a physical effect. Nelson found the REG machines acted like thermometers which measured the dynamics and coherence of the group. There was no effect at business or academic meetings. The machine did not respond to boring meetings. It was only at the moment of like-mindedness when the machine registered order on its normal chaotic non-order.

Nelson was intrigued to see if the REG device would respond to energy at sacred sites. If there were something different like a collective conscious or energy whorl or resonance, perhaps the machine would register it. In America, he observed a medicine man performing a ritual at Devils Tower in Wyoming. At Wounded Knee, South Dakota, he visited an Indian massacre site of the Sioux tribe. The REG machine was affected at both places.

Nelson then took a palm sized REG device to Egypt and visited the Sphinx, the Great Pyramid, the Temples of Karnac and Luxor, plus 23 other sacred sites. His results were surprising. The spiritual energy of the sacred sites registered as high as that of a meditation group. Nelson concluded that both the type of place and activity of the group played a contributing role in creating a kind of group consciousness.

Dr. Roger Nelson founded the Global Consciousness Project, which has accumulated much data regarding his REG machine and consciousness. He found the effects of consciousness do not weaken noticeably with distance. He found that when many minds focus on a single event, the behavior of a worldwide network of REG machines will be altered. Nelson's conclusion was that the collective consciousness of many people could alter the quantum noise around the world. This strongly suggests that the zero point field of the vacuum is affected globally by consciousness.

On September 11, 2001, several hours before the attack on the World Trade Center, the random outputs of REGs in many countries began showing anomalies suggesting an immense global event. The following hours after the 9/11 event, while people were focused on the tragedy, the REGs were affected. One can conclude that if the REG can be affected halfway around the world, it can also affect the local fabric of space. REGs, writes physicist Claude Swanson. REGs can be thought of as objects which tap into the quantum noise arising from zero point energy.

In 1998, Nelson conducted an experiment to measure the "mind of Gaia." He established a network of REGs around the world. Continuous streams of data were sent over the internet to match the moment-to-moment events of modern history. These REG machines were influenced by the death of JFK, Jr., the Clinton impeachment, the bombings of Yugoslavia, the death of Princess Diana and many public ceremonies. However, it did not register the effect with the passing of Mother Theresa or the Monica Lewinsky scandal. The REG strongly suggests there sometimes is a group consciousness.

Morphogenetic Fields

Another pioneer scientist who shed light on the Mind Matrix was Rupert Sheldrake, Ph.D., a biology professor at Cambridge University. He termed this new field of energy the morphogenetic field. It was a mysterious force that connected each individual of a species with all other individuals in its species. Sheldrake suggested that each species had what he called a group mind.

Traditional science asserts that DNA can be thought of as a blueprint. What science does not know is what directs the DNA blueprint to make the decisions that it seems to make. How does a cell know what it should become? Evidence suggests that the organism possesses an interesting property of wholeness. To help explain this dilemma, Sheldrake proposed his theory of the morphogenetic field. He believed the morphogenetic field governs both the structure of living organisms as well as their behavior.

Sheldrake believed that the habits and behavior of any given species in the past builds up through a process he termed "morphic resonance." This morphic resonance affects the habits and behavior of each species. Each species is different and has its own morphogenetic field. In other words, each species has its characteristic morphic resonance. Sheldrake found that morphic resonance allowed members of a species completely separated by geography to absorb new behavioral techniques developed by only a few members of that species.

An example of morphic resonance response was demonstrated by William McDougall who studied rat behavior for 34 years beginning in 1920. He constructed a T-shaped water maze. At one end of the maze he placed a light and at the other end he placed an opaque roof so that it remained dark. He electrified the gangway at the lighted end because he knew rats would want to swim to the light. To McDougall's surprise, he discovered the later generations of rats completed the maze more quickly than preceding generations. He found that the twenty-second generation of rats was figuring out the maze ten times faster than the first generation. He concluded the improvement seemed to manifest in the rat species en mass.

Other researchers around the world began to replicate McDougall's results. They found their rats picked up where McDougal's rats ended. In other words, this knowledge was transferred around the world in the rat species. McDougall also demonstrated that this trait was not passed on genetically.

Sheldrake said that each species has its own morphogenetic field that is intangible but with real spatial structure. The field he found was not static, but allowed a new and beneficial behavior. The behavior learned is not lost when an individual dies. Inborn behavior, habitual behavior, and natural random behavior all work together in dynamic balance. Sheldrake called his theory the Hypothesis of Formative Causation.

Sheldrake claims that DNA contains the blueprint of an organism allowing it to make a conceptual leap. Somehow it is able to tap into this morphogenetic field. This ties into quantum physicist David Bohm's theory which says, "Past forms tend to be repeated or replicated in the present," and that is very similar to what Sheldrake calls a morphogenetic field and morphic resonance. Morphogenetic fields have the ability to span both space and time.

Sheldrake believes that morphogenetic fields might also apply to inorganic substances such as crystals. When a new chemical substance is crystallized, it will always assume its identifying shape. He believes that each crystalline form is controlled by its own morphogenetic field.

Author Michael Talbot believed there is a great social implication with morphogenetic fields. The most obvious implication is that thought and behavior in a substantial number of people is easier for other individuals to tune into, whether beneficial or harmful to the human race. This has come to be known as the Hundredth Monkey Effect based on the behavior of monkeys on one Pacific island washing sweet potatoes being transferred to another group of monkeys on another island. It appears that each species can be thought of as possessing a sort of group mind.

Sheldrake believes that if his morphogenetic field is ever proven correct, it may go a long way toward explaining Carl Jung's concept of the Collective Unconscious and mental telepathy.

Morhogenetic fields, claims Shedrake, are hidden fields which give regular shape and movement to the universe. From the smallest particle, to humans, to galaxies, to all growth and forms, all are determined by actions of morphogenetic fields on matter. These fields serve as channels of subtle energy or blueprints. Sheldrake believes the formation of an atom from a nucleus and electron is guided by one field, the shape of a molecule by another, and the regulation of a cell by another field. Each field interlocks with another, and the field of a larger form orchestrates the morphogenetic field of the smaller forms within it. Sheldrake believes a human is composed of literally billions of fields all directed in an ascending interlocking order up to the general field of that person.

Morphogenetic fields are formed by the very things that are formed. The field is continually being modified. For example, each individual dandelion or species that exists or has ever existed influences the field. For example, the experience of an individual dandelion is transmitted to the field. The field in turn transmits total experience to form every other dandelion. An adjustment a species has to make in order to adapt in one location modifies the field and creates a tendency for a similar adjustment in an entirely different geographic location. This allows adaptation in difficult locations to take place more quickly than could be otherwise in the case of random mutations. The more often that similar individual experiences occur in a species, the more the field is reinforced. This strengthens the field for that species. Laws of nature are in effect habits, reinforced by repetition. Sheldrake believes the laws of nature are not eternal but evolving. The stronger the morphic resonance is, the more likely new generations of dandelions will pick up the vibration and take that form.

This field exists in another dimension. Once a field comes into existence, it does not die. After an entity has reached its final form, the morphogenetic field remains in place and stabilizes the individual against environmental fluctuations. The morphogenetic field helps replace and regenerate lost parts. It directs the growth of an acorn into an oak tree. During reproduction, the sperm and egg create a morphic germ to which a field attaches and begins the process of unfolding an organism. Human behavior is also influenced by morphogenetic fields. New tasks such as driving a car, piano playing, flying a plane, running a four-minute mile, swimming, and taking tests become easier to learn as time goes on.

Sheldrake's morphogenetic field is holographic, and some scientists consider it to be an aspect of the implicate order. Bohm believed there is a guide wave of implicate energy which serves as a formation wave. Both Bohm and Sheldrake's hypothesis lend support to the unity of consciousness.

Consciousness as a whole is a morphogenetic field giving a general shape to each individual consciousness. Each individual consciousness forms its own field that includes its experiences and memories. These individual fields resonate and modify the larger field of human consciousness as a whole, which will affect the future. An individual does have an effect on the whole.

Remote Viewing

Further evidence of a Mind Matrix was provided by two laser physicists, Drs. Harold Puthoff and Russell Targ of the Stanford Research Institute (SRI) in the early 1970s. They discovered another method in which the mind interacts with the level of quantum interconnectedness that they call the information picture. Calling their technique remote viewing, they wrote a number of articles describing their research. Remote viewing is a process where an individual can actually describe what is going on in a distant location when all normal modes of perception are blocked. It is done with the mind.

The research involved choosing a series of target locations that were sealed in an envelope. The envelopes were randomly arranged and stored in a safe. Both the subject and researcher were sequestered. A team of individuals was given one of the envelopes at random and instructed to go to the chosen location specified in the envelope. After 30 minutes of travel time, the researcher asked the subject to describe his impression of the location onto a tape recorder. Constantly, a variety of subjects were able to describe the distant locations where the envelope had been taken. The scientific articles were published in the scientific journal *Nature* and an engineering journal. Over a dozen other studies have replicated the results.

Puthoff and Targ discovered the phenomenon was not limited by distance. The energy involved in remote viewing was not electromagnetic. Some of the best test subjects were people who had little interest in the paranormal, inferring that everyone might have this capability. This suggested to Puthoff and Targ that remote viewing was a latent and widely distributed perceptual ability.

The phenomenon of remote viewing seems to be quite similar to out-of-body experiences. Puthoff and Targ believed it had something to do with quantum interconnectedness. They found at the information level, causality broke down. A surprising development occurred during the research when subjects were able to choose the location even before it had been chosen in the sealed envelope. This suggested to Puthoff and Targ that when the consciousness operates at the level of the information picture, it is no longer bound by space and time. The SRI experiments conclusively proved that humans have the ability to see the future. Other remote viewing experiments have shown that people can accurately describe distant locations even when human observers are not present at the location.

According to Michael Talbot, remote viewing can be looked at as a resonance of meaning conveyed from an object to a mind. He writes, "When harmony of resonance 'meaning' is established, the action works both ways, so that the 'meaning' of the distant system could act on the viewer to produce a kind of psychokinesis that would in effect transmit an image of that system to him (the viewer)."

Both Targ and Puthoff believe that non-local quantum interconnectedness plays a role in precognition. Targ beleives that during a remote viewing experience, the mind appears to be able to access a kind of "holographic soup" or domain. If so, all points are infinitely interconnected both in time and space.

U.S. intelligence agencies have capitalized on this phenomenon of remote viewing. They discovered a big difference in abilities of individuals to perform remote viewing. They found that about one percent of people were capable of consistent, high accuracy required for military usefulness. The CIA was involved with the early remote viewing research but later began to debunk it. The best remote viewers seem to be in a very elite class. Ingo Swann was a very talented remote viewer who had worked with Puthoff and Targ at SRI.

Remote viewer Joe McMoneagle was involved in the military remote viewing program. He stated that accuracy was measured in a comprehensive analysis of his remote viewing sessions. His accuracy was between 65 percent and 80 percent, and it was the same for past, present, and future targets. In one session, he described the building he saw in his remote viewing was actually built many years later. Anecdotes like this strongly suggest that consciousness can easily move in space and time.

At one time, there were a least a dozen remote viewing schools that had been established, many by former military remote viewers. Corporations such as DuPont used remote viewing techniques to help scientists to become more creative. In medicine, some doctors have been trained to remote view inside their patients to better diagnose illnesses. The Chinese government has tried to identify talented remote viewers in China. In 1984, they researched an eleven-year-old boy named Xie Zhaohui. The boy was asked to diagnose 105 patients with various illnesses. After gazing at the patient for several minutes, he diagnosed correctly 93 of the patients and misdiagnosed 12 of them. The U.S. intelligence agencies have used remote viewers to predict future attacks and even tried to find Osama Bin Laden, as reported in the *Sunday Times* of London. Remote viewing does support the existence of some type of Mind Matrix.

Transcendental Meditation

As we have discussed earlier, scientists are finding that when a group of like-minded individuals focus on a similar thought, the effect detected goes way beyond the group itself and way beyond the locale of the group. This suggests that a small group of people can effect a large population, which was proven with Transendental Meditation (TM) research, showing that consciousness of the Mind Matrix can greatly influence events and actions.

TM research was conducted in 24 U.S. cities with at least 10,000 population. A 1972 experiment found a meaningful change in the community when as few as one percent of the population participated. The experiment involved using specific meditation techniques to create the inner experience of peace that was transferred to the world around them. This was termed the "Maharishi Effect" in honor of the yogi Maharishi Mahesh. The Maharishi stated that when one percent of a population practiced his method of meditation, there would be a reduction in crime and violence for that population. Overall, the cities that had one percent of the population practicing TM mediation averaged a crime rate decline of 22 percent compared to a 2 percent rise in the control cities.

In a follow-up study of 48 cities that participated in the research, half of which had a one percent TM meditation group meditating, there was an 89 percent reduction in the crime trend, compared with a 53 percent increase in the control cities.

A 1993 TM two-month study was conducted in Washingon D.C. called the "National Demonstration Project." When the local Super Radiance Group, who praciced TM meditation, increased to 4,000 people, violent crime, which had been increasing, fell 24 percent until the end of the experiment. As soon as the group discontinued, the crime rate increased again.

The TM organization studied the effects of group meditation

on world peace. A 1983 study, called the International Peace Project in the Middle East, was published in the *Journal of Conflict Resolution* in 1988. The study took place in Israel where a group of TM meditators tracked the Arab-Israeli conflict day by day for two months. Practitioners were trained in precise techniques of TM to create peace in their bodies, rather than simply thinking about it in their minds or praying for it to occur. On days when the number of meditating participation was high, war deaths fell by 76 percent in Lebanon, and local crime, traffic accidents, and fires also decreased.

The study showed that when a small percentage of the population achieved peace within themselves, peace was reflected in the world around them. The researchers calculated the minimum number of people to experience inner peace before it was mirrored in the surrounding world. They found it would take the square root of one percent of the population representing the base number for the effect to begin. The more people participating, the more pronounced the effect.

The study was able to determine a formula for peace to ascertain how many people were needed to work together for peace and healing. The formula was based on the number of people living in a given locale and then calculating the square root of one percent of that population. For example, in a city of one million people, the total of TMers needed was only 100 people to bring about a change in peace. For the planet of six billion, it would only take about 9,000 people to have an effect, the minimum number to begin the process. The more people, the better the effect. These numbers are small enough to have a great potential in making this a more peaceful planet on which to live.

DNA and the Mind

Science is beginning to demonstrate that DNA is also connected to this Mind Matrix. In 1995, Vladimir Poponin and colleagues at the Russian Academy of Sciences performed a number of experiments that suggested human DNA directly affected the physical world through what they hypothesized to be a new field of energy connecting the two. Poponin wrote,

"We believe this discovery has tremendous significance for the explanation and deeper understanding of the mechanisms underlying subtle energy phenomenon, including many of the observed alternative healing phenonmenon." He called his experiment the "DNA Phantom Effect."

Poponin's experiments tested the behavior of DNA with light particles (photons) of which the universe is made. The researchers created a vacuum by removing all the air from a specially designed tube. Photons were introduced into the tube and distributed in an unordered manner. They were then located and measured. Then a DNA sample was placed inside the closed tube. Once the DNA was placed in the vacuum, the photons arranged themselves differently in the presence of the living DNA. The DNA shaped the photons into a regular pattern. The scientists concluded that the DNA had a direct effect on the photons. After removing the DNA, the scientists were surprised when the photons remained ordered, just as if the DNA were still in the tube. The remaining question was what was the force that affected light particles and connected it with DNA? Poponin concluded, "We were forced to accept the working hypothesis that some new field structure is being excited." The new field sounded familiar to Max Planck's Mind Matrix that he identified a half-century before. The experiment showed a direct relationship between DNA and the energy our world is made of, and which had changed the behavior of light particles.

Gregg Braden writes about another experiment that took place in the 1990s regarding how emotion influences the cells of our body. Scientists working with the U.S. Army researched the power of feeling and its effect on living DNA cells removed from the body. They wanted to find out if emotions still impacted cells once removed from the body. The study was published in 1993 in the journal *Advances*.

A DNA sample was collected with a swab from inside a subject's mouth. It was then isolated and taken to another room separate from the subject. Electrically, the DNA was measured to see if it responded to the emotional changes of a volunteer, located several hundred feet away. The donor underwent a spectrum of real emotions for a brief period of time and the DNA was monitored in another room. At the same time of the volunteer's emotion, the DNA showed a powerful electric response, acting as if it were still physically connected to the body. Researcher Cleve Backster had designed this experiment for the Army and was able to replicate this experiment up to 350 miles. Every time the donor experienced an emotion, the DNA reacted. This strongly suggested there was a quantum field, Mind Matrix, connecting all matter. The experiment concluded that emotion has a direct influence on DNA and the cells communicate through this field of energy regardless of distance.

Backster is also a well known polygraph researcher. In one experiment he hooked up the polygraph to a plant. In another room was a boiling pot of water. As live shrimp were being introduced into the boiling water, the polygraph of the plant went wild. The plant was picking up the emotion of the shrimp.

Further research at the Heartmath Institute by Glen Rein and Rollin McCraty found that human emotion could change the shape of DNA without touching the DNA. Subjects created precise feelings and were able to influence their DNA located in a beaker. The researchers discovered that different intentions produced different effects on the DNA molecule, causing it to either wind tighter or unwind.

From the different DNA experiments described, one can draw several conclusions as expressed by Gregg Braden in his book *The Divine Matrix:*

1. There is a matrix of energy that connects one thing with everything else in the universe.
2. Our body's DNA gives us access to this energy matrix connecting the universe. Emotion appears to be a key to tap into this matrix.

This energy matrix is the Mind Matrix, the quantum, field, the implicate order, or the Tao. It is everywhere and is everything. It makes up all space. As creators, we are connected to this Mind Matrix and are participants in the constant change that helps provide meaning to life.

Chapter Six

SUBTLE ENERGY

Life Force of the Universe

In previous chapters we have alluded to the life force energy that penetrates the universe in the void or zero point field, referred to as zero point energy in the scientific arena. In Eastern terminology it is called chi (qi) or prana. In Western terms it is called subtle energy. This life force energy underlies all living and nonliving aspects of the universe. Subtle energy is the bridge between science and consciousness, because subtle energy responds and interacts with consciousness. Subtle energy affects everything in the universe including gravity, time, mass, water, electromagnetism, light, vacuum, nuclear, dark matter, and living physiology. This energy has been studied for thousands of years by spiritual masters such as shamans, yogis, and Qigong masters. What has kept Western science ignorant about subtle energy is the lack of devices for measuring this energy. Western science is based on measurement. However, one man, Dr. Claude Swanson, has gathered the most comprehensive science underlying subtle energy in his landmark book *Life Force: The Scientific Basis*. Swanson was graduated from MIT and received a Ph.D. from Princeton University in gravitational physics. Much of this chapter is based on Swanson's writings and research of subtle energy, more evidence of the Mind Matrix.

Chi and Prana

Chi energy in China and prana in India are considered one and the same. This energy, which constitutes a fifth force in China, has been ignored by Western physics. Chi energy can explain many of the phenomena that Western science cannot, such as paranormal events and long distance healing. Esoteric literature has long claimed that prana is a particle emitted and absorbed by all living things. Evidence of this energy can also be found in the Earth's ley lines, in Kirlian photography, auras, acupuncture meridians, and alternative healing.

Ley Lines

For more than 4,000 years, ancient texts have talked about chi energy flowing through the landscape. In ancient Europe, this energy flowed along ley lines, in straight lines often marked by megaliths and sacred sites, such as Stonehenge and Avebury. Ley lines of the Earth are similar to the acupuncture meridians of the body through which chi energy flows.

Ley lines are found all over the world. In Bolivia, ley lines connect the sacred sites of the Aymava Indians. The Romans built roads over the ley lines, often over routes existing long before the Romans arrived. When the Spanish arrived in Peru, they followed the roads built on ley lines to the seat of government of indigenous leaders. The Aborigines in Australia called them "song lines," which carried the life force energy to make plants and animals grow.

In China they call them "dragon lines," and by law only temples could be built upon them. Legends in both England and China say that ghosts and spirits travel along these lines. The Chinese art of feng shui is devoted to managing and controlling chi energy by engineering and landscape design.

Dr. Swanson writes that sacred sites are locations where subtle energy accumulates. These sites have been designed to maximize the amount and type of energy which collect there. Often they are situated where ley lines cross, giving rise to subtle energy vortices. They are often associated with paramagnetic rock, which stores the positive polarity of the yang chi energy. Ley lines carry either yin or yang energy, each having different effects. Yin energy is calming and healing while yang energy gives a feeling of power and promotes energy and activity. Locations of ley lines can be identified by dowsing.

Kirlian Photography

Subtle energy can also be detected by the use of Kirlian photography. Kirlian photography is a short pulse of high frequency electromagnetic voltage applied between the subject being photographed and a metal plate. The theory is that when chi energy is rapidly accelerated by a high frequency electric field, it emits light. Kirlian images show detailed features providing information about a subject's physiology.

Kirlian light primarily appears at known acupuncture points. It is also known to vary with the emotional state of the subject. Harry Oldfield discovered that food which has been processed, frozen, or had pesticides on it showed a diminished Kirlian aura.

Researchers have discovered energy patterns via Kirlian photography where parts of the plant had been cut away. The image of the missing leaf is called the "phantom leaf effect," meaning if a leaf had been cut from a plant, it would still show up on a Kirlian photograph. This suggested that every living creature is surrounded by an energy field which carries a pattern or template for the physical object, much like Sheldrake's morphogenetic field.

Auras and Chakras

Auras and chakras are another good example of subtle energy. Western science refers to this light surrounding a body as the "bio-energy field." In Chinese and Hindu medicine, an aura is defined as an energy field surrounding the body. It can even be seen by some people. It is closely involved in life processes. A number of Russian and Eastern European researchers have been able to photograph the aura in the ultaviolet region of wavelengths. Dr. Walter Kilner viewed the aura of more than sixty patients without a prior physical exam. He found that irregularities in their aura correlated with physical health problems. According to Chinese medicine, the chakra is defined as a spinning wheel of subtle energy, one of a number of vortices in the human aura through which energy flows.

Harry Oldfield developed a camera called the Polycontrast Interference Photograph (PIP) camera which uses software that works with a digital video camera. It produces images which show the subtle energy fields around people and animals. It has been successfully used to identify health problems.

In Russia, Dr. Konstantin Korotkov developed a device which takes Kirlian images of the fingertips, which in turn converts them into a virtual image of the aura of the entire body. He called his invention the Gas Discharge Visualization (GDV) device. It is virtual because the whole body aura image is computer generated using signals from each fingertip. There are many acupoints on the fingertips that are connected by acupuncture meridians with the body's organs. Clinical trials since 1991 have found a correlation of up to 98 percent between the GRV diagnosis and conventional diagnostic technique. The GRV also revealed the aura changes in different conscious states.

Acupuncture

Research has shown that acupuncture meridians are universal, playing an essential role in the growth and regulation of all life forms. Acupuncture meridians serve as a path for chi energy to flow, enabling cells and organs to communicate. They produce and maintain a coherent, holographic pattern which guides growth and healing of the organism. Oriental medicine for centuries has taught that acupuncture meridians control and distribute chi, the life force energy. Its signals are believed to arise directly from the DNA, signals created by biophotons to be discussed shortly. Acupuncture meridians carry these signals to all parts of the body where they are read and decoded by the cell's DNA.

Swanson reports that recent studies have been able to identify the existence of acupuncture meridians in physical organs. These meridians are sometimes called the Bongham duct system. Radioactive phosphorous is injected into the acupuncture system. It does not appear in the blood or lymphatic system, but only at acupoints and acupuncture meridians. The research by Dr. Kim Bongham showed there is an independent system of tubules involved in acupuncture, which carries a slow flow of material within the network.

Dr. Bongham tested many organisms and concluded that all multicellular creatures, including all plants and animals, possess an acupuncture meridian system. He found that this system plays an important role in morphological differentiations and in the growth of the organism. It carries the shape and structure information of the body. His research confirmed there is a correlation between meridians and the organs to which they connect.

Traditional science is beginning to catch up with ancient knowledge. A small volt-ohm meter can verify the existence of acupuncture points. As an acupuncture point is touched, the conductivity suddenly increases as measured by the meter. Each acupoint is connected to a specific meridian and organ, forming a complex network. The circuit energy at each acupoint can be measured with the volt-ohm meter to aid in diagnosis and treatment.

For centuries, Traditional Chinese Medicine (TCM) has asserted that blockage of the acupuncture meridians prevents the proper flow of chi energy to various organs, resulting in health problems. Stimulating the acupoints will eliminate the blockages, allowing the proper flow of chi energy. There are six ways to stimulate acupoints and restore proper chi flow. They include:

1. Chemical Stimulation

2. Manual Massage
3. Acupuncture Needles
4. Electrical Currents
5. Laser Beams
6. Psychic Energy or Energy Healing

There are twelve primary meridians that either start or end in the fingers or toes. The success of acupuncture in promoting healing is just one more piece of evidence of this life force energy.

Electro-acupuncture was developed by Dr. Reinhold Voll, a German physician, who developed the Voll Dermatron. The Dermatron measures electrical conductance to evaluate health. Voll's research was based on earlier work by Walter Schmidt, who found that acupuncture points associated with diseased organs showed higher electrical resistance than healthy organs. The Dermatron uses a probe that passes a small current, around 10 microamps, into the skin at a test acupoint, usually a fingertip. The other end of the circuit is completed by a brass ball or cylinder held in the subject's other hand. Voll's research used healthy individuals as controls and compared their findings to subjects with specific illnesses. He established standard values which could be used to assist in diagnosis. A low conductivity reading less than the established threshold indicated a health problem, with the acupoint measured correlating to its corresponding organ. The Dermatron can also assist in determining a homeopathic remedy to facilitate the flow of chi to heal the condition.

Subtle Energy Healing

Giving evidence to the nonlocal aspects of chi energy is Qigong, which is part of Traditional Chinese Medicine. Qigong masters can project their healing energy over thousands of miles, achieving great success in both diagnosis and cures. Qigong is similar to pranic healing in which subtle energy is transmitted to a recipient for healing. Professor John Neely, M.D., at Penn State University, conducted a long distance experiment with

Qigong master, Jixing Li. Neely repeatedly showed the distant healing chi energy from Li was able to disrupt cancer growth without affecting nearby cells.

In India, energy healing is called pranic healing where prana is sent from a healer to the diseased cells of a sick person. Like Qigong, pranic healing can take place from a long distance. Scientific studies have found significant results in pranic healing independent of distance. Studies also found that healing was just as effective thousands of miles away as it was near. In all cases, pranic healing increased cell survival rates in specific experiments. Both pranic healing and Qigong go back millennia in India and China.

Experiments have also shown that energy healing can be effective against cancer, even when the senders and recipients are not believers. This suggests there is some objective energy field occurring in the healing process.

Research by Dr. Joie Jones discovered that subtle energy can also exist in negative forms that can hinder energy healing. Removing this negative energy through various techniques allows successful subtle energy healing.

Scientists are beginning to understand how subtle energy works with much of the research being done in Russia on biophotons and torsion.

Torsion and Time

For many decades Russian scientists have been conducting breakthrough research helping to understand the properties of subtle energy which they named torsion. They also can explain scientifically how torsion affects the flow of time. Torsion (chi) is composed of two different forms of energy, yin and yang. One form of torsion twists to the right and the other to the left. They discovered these two energies have opposite effects on the flow of time and energy, resembling yin and yang.

Torsion (chi) is produced by the interaction between elementary particle spins which generate a twisting of space. A number of scientific papers have been published by the Russians about torsion, indicating that it is exactly the same as subtle energy. Russian research has shown that torsion travels faster than light and has the ability to alter both mass and time. Some torsion devices have affected consciousness.

Research has shown that torsion energy has two polarities, left and right particle spin. This subtle energy of chi appears to decrease entropy (the degradation of the universe) and promotes order within a system. It can slow down the flow of time, and it can assist in the body's healing. In Russia, this subtle energy has even reached the level of engineering used to alter metal crystallization, assist in mineral prospecting, and provide a new form of communication.

Each elementary particle (such as an electron, proton, or neutron) exists as a result of a complex interaction with all matter and energy in the universe, writes physicist Dr. Claude Swanson, including matter and energy that exist in both the past and the future. The resulting radiation exchange with all these particles makes up the pattern of energy around each particle. The energy pattern is a four-dimensional hologram in space and time.

The energy of the vacuum (zero point field) has a structure that is preserved and maintained by each particle. According to Swanson, matter cannot exist without the supporting holographic pattern in the vacuum energy. An additional form of energy was discovered by the Russians, which they called the "torsion field." It is created whenever photons are created, but can persist in the vacuum and carry form and structural information that affects other physical processes. This torsion field, according to Swanson, is believed to carry the fundamental holographic pattern of information which guides growth. This sounds like Sheldrake's morphogenetic field. It provides a physical basis for explaining auras and long distance healing.

Nearly a half-century ago, Dr. Nikolai Kozyrev, a Russian astrophysicist, discovered a new force in physics he termed the "density of time." Kozyrev concluded that the rate at which time passes can be altered by other physical processes such as a twisting effect in space by torsion. Torsion interacts with spins of all particles, including all elementary subatomic particles such as electrons and protons. Kozyrev's torsion theory and time density theory explains that when entropy increases in one place, it will decrease elsewhere. Torsion can be stored in an object and can also be transferred from one object to another. Russian scientists believe this torsion energy is subtle energy or prana or chi.

Time density implies that space must have a twisting property. As particles accelerate, they must produce a twisting effect in space because of this spin. Whenever a charged particle produces an electromagnetic wave or photon, it also creates a torsion wave. At the point of absorption of a photon, negative entropy is produced, balancing the increased entropy at its source.

Scientists have discovered that torsion waves can travel backward in time from the future and forward in time from the past. A combination of the two waves generates nonlocal torsion waves that travel much faster than light.

Kozyrev discovered that a field of flowers creates abundant torsion (chi) in the springtime. He also found sunshine creates left-handed torsion, confirming other scientists' research. Both examples decrease entropy and absorb time. On the other hand, a dying plant is associated with right-handed torsion associated with increasing entropy.

Kozyrev says in the physical process, the cause of an event and the effect are always separated by a small amount of space and time. The cause is an initial imbalance of force in space (ripple) and the effect is the response that tries to bring it back into balance. The cause and effect dance is separated in space and time by a small distance, which has the properties of a twisting vector that alters time. He found that the right-hand twist of the particle moves time forward. When entropy increases at one point because of an irreversible process, the local time density is altered, creating a right-handed torsion wave. At a later time, the torsion will be absorbed, resulting in an increase in time density and decrease in entropy at that location. Torsion waves are generated whenever a particle changes directions or turns a corner or zig zags, referred to as a "vertex." These waves are able to travel backward or forward in time.

The torsion wave maintains overall balance in entropy for the entire system. Torsion waves are nonlocal, so any process involves the entire system of the universe. There are no isolated systems. Both torsion and time density permeate the entire universe and connect any local process to every other particle in the universe. This science explains how the universe is connected. Everything is coupled. One process affects every other process in the universe, including the flow of time. Kozyrev wrote, "Time carries order of negative entropy and it is either emitted by a system when its order decreases, or absorbed by a system when its order increases."

Every elementary particle with mass has a spin, and each one generates its own torsion field. Each spinning electron also creates a small magnetic field.

Swanson proposes that the Russian model of torsion operates at a deeper level than quantum physics. It relates to the underlying fabric of space occurring on a very small scale, less than one billionth the size of a proton. The vacuum can be thought of as a sea of energy comprised mostly of created particles that pass through it, including protons and torsion waves.

Torsion can be viewed as a flux of left-handed and right-handed massless particles within the vacuum. The relative number of particles with each spin direction determines the rate of passage of time. These particles make up electromagnetism and gravity. Right-handed torsion originates from entropy and repels mass. Left-handed torsion attracts mass. According to the Russians, torsion is the source of dark matter in space that enhances gravity. The vacuum is filled with massless, one-half particle spins, which carry the torsion force called "phitons" that carry left and right spins of particles. Together they make up the fabric of space-time. Left-handed spin particles push forward in time and right-handed waves push backward in time.

Torsion waves always travel at the speed of light and can still produce nonlocal effects at distant locations. It can do it because of the two waves – one traveling forward in time and the other backward in time. The torsion field also has components that can travel at infinite speed because of these two waves.

Every event creates torsion, and space is filled with event patterns. It is possible to tap into this web and read history. Swanson has proposed that maybe this is the physical origin of the Akashic Record, the storehouse of all our past life memories and history.

The torsion field can also be shaped by consciousness. A particle responding to torsion in the zero point field is buffeted forward or backward in time. Swanson writes, "To be awake means realizing that our minds tune into a parallel system of reality and synchronizing with it. In doing so, we select the synchronized plane we accept as real." Perhaps torsion can explain parallel universes.

Biophotons

Biophotons were discovered in the 1930s by another Russian scientist named Alexander Gurvich. He discovered that the growth on the root of an onion affected the growth of a separate onion root. Through a series of experiments, he showed one onion was communicating with the other onion by weak light signals.

Vlail Kaznacheyev, M.D., also of Russia, discovered that light from cells could also cause death. His experiment involved two petri dishes of bacteria. He introduced a poisonous virus into one dish, killing the bacteria. The bacteria in the unconnected dish also died. He concluded that"... an ultra-weak electromagnetic radiation emitted by living cells is a carrier of information about cell disease and death from one culture to another."

Much of the knowledge about biophotons comes from the research of Dr. Fritz Albert-Popp of the International Institute of Biophysics in Germany. He discovered that pulsed light was created and absorbed by the molecule's DNA. He also found that biophotons from the body created a hologram.

In quantum physics, light occurs in small packets of energy called photons, which are also referred to as particles. **A photon is the smallest amount of light that can be emitted by an atom or molecule. It is generated when an atom or molecule undergoes a change of state, as in shifting to a state of excitation.** The human body emits photons at the rate of a few to a few hundred per second from one square centimeter of skin. In living organisms, certain types of photons are emitted and received by DNA as well as a few large molecules. DNA biophotons are mainly created in the cell nucleus. Functions of biophotons include biocommunications, stimulation of biochemical reactions, and coordination of biochemical events within the body. Some scientists believe they may play a more important role than biochemistry.

Biophotons enable chemical reactions to build complex life forms. The waves of energy from biophotons create an interference pattern that produces a holographic field of energy around the body, like a morphogenetic field. Swanson proposes the hologram is read by the DNA and provides the blueprint for cells, organs, and tissues. DNA molecules throughout the body create this field and in turn are controlled by it.

Both biophotons and torsion are created together. Torsion fields are created whenever the spin of subatomic particles changes. The creation of a photon occurs with the acceleration of an electron that changes its spin, producing a torsion wave with its creation. Torsion fields are believed to radiate a long distance. They carry the holographic information about the body which can explain long distance healing and other consciousness phenomena.

Biophotons in the acupuncture meridians create holograms which give patterns for organs and tissues that are read by the DNA to maintain structure of the body. Similarly, the torsion field creates holographic patterns as well, both inside and outside the body. These torsion fields are responsible for the auras, chakras, aspects of the acupuncture system, and the flow of subtle energy within the body. Acupuncture meridians have been found capable of amplifying and retransmitting biophotons throughout the body. Studies have shown that biophotons can pass through many cells before being completely absorbed and reflected. Thousands of cells come in contact with the same photon. They become synchronized, and their phases become aligned into coherence.

A number of studies have shown that biophotons can even be photographed outside the body, but biophotons found outside the body suggest an injury or illness. Healthy organisms maintain large stores of trapped energy inside the body.

When biophotons are phase-conjugated, they generate torsion waves that propagate beyond the body. They form a holographic pattern around the body. This field interacts primarily with the particle spin and contains very little energy. It maintains the holographic information outside the body according to Swanson. It is the torsion field, which is the physical basis for the aura.

It is the biophoton field in the body that generates the torsion waves every time a biophoton is created. These waves are created by the changing spin of the electrons of the biophoton. It is the torsion waves, not the biophotons, that travel from the healer to the patient. When torsion fields are coupled with the biophotons, they carry the same information about the health of the person that travels much faster and requires little or no energy.

Something does escape from inside the body and that is called torsion. Biophotons are internally reflected within the body, keeping inside most of the energy needed for biochemical reaction.

Every human is surrounded by a torsion field that is maintained by its biophotons. Through chromatin, the biophoton is coupled directly to the DNA. Each individual seems to have a unique set of frequencies of biophotons because each person's DNA is unique. This can explain why healers can tune in to particular individuals at a distance. Individual unique frequencies may also relate to the identity of a soul and its ability to remember past lives, writes Swanson. The concept of the torsion field provides a physical mechanism for the Vedic concept of the Akashic Record.

Biophotons have shown us that the essential nature of life can be found in this coherent system of energy created by the DNA. The body's biophotons at the physical level are coupled to maintain the hologram of the physical body. Biophotons are the basis on which subtle energy acts on the body to maintain the life force of living organisms.

Chapter Seven

ACCESSING THE COLLECTIVE CONSCIOUS

At Our Fingertips

The Mind Matrix (quantum field) is all around us carrying information about the past, present, and future. By tapping into this subtle energy we could be omniscient and all knowing. The potential is enormous if there were a technique to access this infinite knowledge. If we are all connected in the universe as we have described, there must be an avenue to access this energy of information that is all-encompassing – the implicate order. Do not despair, as there are techniques that have been developed allowing us to access the collective conscious. They include applied kinesiology, dowsing, and radionics. David R. Hawkins, M.D., Ph.D., has provided some of the best research to tap into this wealth of information in his book *Power Vs. Force.*

Applied Kinesiology

Muscle testing is a term used to describe applied kinesiology, a technique pioneered by chiropractor Dr. George Goodheart. He discovered that the strength of every muscle is connected to the health and pathology of a corresponding body organ, and he also found that each muscle is associated with an acupuncture meridian. His most important finding was that muscles instantly become weaker when the body is exposed to a harmful stimulus and stronger with a therapeutic substance.

Muscle testing is a very simple technique. The subject

extends one arm out straight to the side at shoulder height and is asked to provide resistence when the practitioner pushes the arm down. At some level below conceptual consciousness, the body knows what is good or bad or true or false and that knowing is registered in the muscles. The practitioner asks a yes or no question and applies pressure to the extended arm. If the body's answer is "no" or is harmful to the subject, the muscles become weak and the arm is easily pressed downward. If the answer is "yes" or beneficial to the subject, the arm remains firmly in place.

In the 1970s, psychiatrist John Diamond, M.D., refined muscle testing into a new discipline he called behavioral kinesiology. He taught his method to thousands of professionals at seminars around the world. Diamond discovered indicator muscles would either strengthen or weaken in the presence of a positive or negative emotional, physical, or intellectual stimuli. Diamond's research found uniformity of response among his subjects. His results were predictable, repeatable, and universal. Diamond found the body responded accurately, even when the conscious mind was unaware.

For example, he found certain abstract symbols caused all subjects to become weak, and other symbols would have the opposite effect. All classical music and most pop music caused a universally strong response. Hard rock music produced a weak response. He found that subjects listening to tapes that contain falsehoods (even though the speaker seemed to tell the truth) produced a weak response in subjects. Being able to test for truth intrigued Dr. David Hawkins, a renowned psychiatrist in New York City who now lives in Sedona, Arizona. In 1975, Hawkins began his kinesiology research in responses to truth and falsehoods. His research results and spiritual journey are described in his book *Power Vs. Force.* (All quotes from Dr. Hawkins come from this book.)

Dr. Hawkins performed double blind studies in mass demonstrations that involved envelopes that contained artificial sweeteners. He found subjects universally tested weak in response to unmarked envelopes containing artificial sweeteners and strong to identical placebo sweeteners. Hawkins believed something was at work in a form of communal consciousness. He found that questions must be phrased so that the answer is very clearly "yes" or "no." He believed at some primal level, the brain is connected to a universal energy field.

The next phase of Hawkins' research involved a calibration scale of relative truth by which intellectual positions, statements, or ideologies could be rated on a range of 1 to 1,000. The point of common weak response on a group of subjects determined the calibration. Ideological validity could be appraised as an innate quality in any subject. In other words, Dr. Hawkins found test subjects would come up with the same response to a "yes" and "no" question. For twenty years, Hawkins analyzed the full spectrum of the level of human consciousness.

Hawkins' research showed that the anatomy of consciousness produces a profile of the human condition. It allowed a comprehensive analysis of the emotional and spiritual development of individuals, societies, and the human race in general. It provided an understanding of man's evolvement in the universe. His calibration scale serves as a guide to where we are on the ladder of spiritual enlightenment. For the first time in Western science, Hawkins confirmed what mystics and sages have always said about self, God, and the nature of reality – the Infinite Oneness of all. Following a serious illness, Hawkins had realized there was no personal self-ego left, just an Infinite Presence of such unlimited power that it was all that was. His world was illuminated by his clarity of an Infinite Oneness, which expressed itself as all things.

Hawkins' research and training in psychiatry found that the origin of emotional sickness lay in people's belief that they were their personalities. Upon Hawkins' spiritual awakening, he discovered that there are no "events" or "things," and that nothing actually "happens" because past, present, and future are only artifacts of perception, as is the illusion of a separate "I" that is subject to birth and death. Hawkins believed that it is only the illusion of individuality that is the origin of all suffering. When one realizes he or she is one with the universe and all that is, there is no further suffering. When Hawkins' false self dissolved into the universal self of his origin, he experienced a state of absolute peace and relief from all suffering. Hawkins said that all pain and suffering arise solely from the ego and not from God.

Dr. Hawkins saw kinesiology as a "wormhole" between two universes – the physical and the mind and spirit – an interface between dimensions. Kinesiology provided a tool to reconnect with the higher reality and demonstrate for all to see, said Hawkins. Kinesiology is now a well established science based on the testing of an all or none muscle response stimulus. It provides for the first time in human history, writes Hawkins, an objective basis for distinguishing truth from falsehood. Hawkins' technique can be used to calibrate human levels of consciousness using an arbitrary logarithmic scale of whole numbers.

Hawkins said, "The individual human is like a computer terminal connected to a giant database. The database is human consciousness, of which cognizance is merely an individual expression, but with its roots in the common consciousness of all mankind. The database is the realm of genius: because to be human is to participate in the database, everyone by virtue of his birth has access to genius. The unlimited information contained in the database has now been shown to be readily available to anyone in a few seconds, at any time and any place . . ." The database Dr. Hawkins talks about transcends time, space, and any limitations of individual consciousness. Hawkins' technique can be used by anyone, anywhere, and anytime.

Kinesiology demonstrated for the first time the interconnection between mind and body, revealing that the mind thinks with the body itself. It has the ability to explore how consciousness reveals itself in the mechanism behind the disease process. Hawkins said the last great barrier to human knowledge is the investigation into the nature of consciousness. However, Hawkins' research has indeed shown that consciousness can be investigated because everything in the universe is connected with everything else. Quantum physics has shown that the subjective and the objective are one. If the subject and object are the same, time can be transcended. If they are one, then we can find the answers by merely looking within man himself, said Hawkins. The body can discern to the finest degree the differences between that which supports life and that which does not.

Hawkins' research supports Rupert Sheldrake's morphogenetic field theory hypothesis and Karl Pribram's holographic model of brain-mind function. His research spanned twenty years involving millions of calibrations on thousands of randomly selected subjects of all races, sexes, and ages. Without exception, said Hawkins, the results were identical and entirely reproducible. His research method was used to verify the truth or falsehood of a declarative statement. He discovered that no reliable results could be obtained by inquiring into the future. Only statements regarding current or past events would produce consistent results.

In developing a map of human consciousness, Hawkins called the process "attractor research." Attractor is the name given to an identifiable pattern that emerges from a seemingly unmeaningful mass of data. Attractor patterns represent energy fields that are qualities of consciousness itself. The calibrated levels of consciousness represent powerful attractor fields within the domain of consciousness itself. Each consciousness level defines content, meaning, and value specific to that level of consciousness, on a scale of 1 to 1,000 that represents the widespread pattern of human consciousness and behavior. In this way Hawkins identified power ranges of energy fields.

Hawkins' Reasearch

For many years Dr. Hawkins conducted research on the consciousness of humanity. Hawkins' research concerned itself with a limited set of parameters of consciousness that was calibrated from 1 to 1,000, representing the logarithm (1 to the base 10). The range 1 to 600 represents the vast majority of human experiences. From 600 to 1,000 is the range of nonordinary evolution of enlightenment where sages reach the highest spiritual state. His goal was to generate a practical map of the energy field of consciousness. He did this using kinesiology on thousands of subjects by asking yes and no questions.

Hawkins found the critical point in his data was in the range of 200. This level was associated with integrity and courage. Below 200, the power of attractors was described as weak or negative. Above 200, they were strong or positive. Below 200, the primary impetus in life was to survive, which included attributes of anger, pride, and shame, which was the lowest. These traits are characteristic of egocentric impulses. When consciousness rises above 200, the well- being of others becomes increasingly important to the subject. At the level of 500, the happiness of others emerges as the essential motivational form. At the 500 level, the person becomes interested in spiritual awareness for both oneself and others. At the 600 level, the subject is motivated to work for the good of mankind and to search for enlightenment. Scores from 700 to 1,000 indicate a dedication to the salvation of humanity. Hawkins found that subjects reacted similarly to different attractors around the world. There is an enormously powerful attractor field organizing all human behavior. Interaction of variations within the attractor fields make up the history of civilization and mankind.

Hawkins' map of consciousness transforms one's understanding between causality and evolving consciousness. It becomes apparent, according to Hawkins, that by taking responsibility for the consequences of one's own perception, the subject can transcend his or her view from the role of victim to an understanding that nothing out there has power over him or her. It is not life events, but how you react to them and your attitude that determines whether events have a positive or negative effect on one's life. Hawkins emphasizes that nothing outside one's perception has the power to create stress. Where force meets power, like a good attitude, force is eventually defeated.

All levels below 200 are destructive of life, including both for individuals and society at large. Levels of consciousness are always mixed. In one area, a person may operate at a different consciousness dissimilar from another area. An overall individual consciousness level is the average effect of the various levels. For many centuries the collective conscious was at 190 as determined by applied kinesiology, but around 1990 it had risen to 207. Only 0.4 percent of the world's population has reached the level of love at 500 on the scale, and only 1 in 10 million reach the level of peace at level of 600. Einstein and Freud were calibrated at 495.

Enlightenment is at the level of 700–1,000. At this level is Oneness. There is no longer the experience of personal self separate from others. Hawkins says at this level the unmanifest is experienced as self beyond mind. This is the level of non duality. The ego has been transcended. Consciousness is not localized because awareness is present everywhere. Some examples of those who have reached this level are Jesus, Buddha, and Krishna.

Hawkins found that 85 percent of humanity were calibrated below the critical level of 200. Many people who live in poverty calibrate at the apathy level of 50. Much of the remainder of the world live in fear at 100. The majority of people live their lives in quest for security.

Hawkins said it is uncommon for people to progress from one level to another during their lives. On the average, the level at birth increases an average of five points during the lifetime. Motivation is what defines one's level, and motivation proceeds from meaning. He says that 2.6 percent of the population accounts for 72 percent of society's problems.

Hawkins subscribes to Bohm's theory of a holographic universe. Kinesiology reveals the hidden implicate order by making it explicit. The enfolded universe, the implicate order, is connected with human consciousness. Bohm had said that meaning links mind and matter.

Carl Jung calls this hidden order the collective unconscious. Hawkins describes the collective unconscious as "a vast database of human awareness which is characterized by powerful, universal organizing patterns. Such a database is comprised of all of the information ever available to human consciousness." He describes it as a giant storehouse awaiting a retrieval process. This great database gives us the potential to know virtually anything the moment it is asked for. It is the storehouse of everything that has been experienced anywhere in time. This is the origin of all information attained by intuition, premonition, divination, or dreams. This is the energy field that kinesiology taps into. This is the Mind Matrix.

Hawkins writes that omniscience is both omnipotent and omnipresent. There is no distance between the unknown and known. Merely by asking, the known is manifest from the unknown. Human consciousness can transform an unseen concept into a manifest experience – the implicate into the explicate. The ultimate reality, writes Hawkins, is existence without form. This includes yin and yang, the manifest and unmanifest, the form and formless, and the seen and unseen.

Pure Consciousness

The nature of consciousness has never been clinically examined, states Hawkins, with Western medicine believing it is only a function of the brain. The French philosopher Rene Descartes believed that consciousness is only aware of itself when it assumes form. Mystics claim that consciousness is beyond form, a theory supported by David Bohm's implicate order. Without consciousness there would be nothing to experience form. Form becomes an experience of the formless. In metaphysical terms, consciousness is an act of beingness implicit to man's definition of himself. A person identifies with the body because the mind is experiencing the body. Consciousness itself resonates with both awareness and experiencing, both of which are subjective.

Hawkins said one characteristic of pure consciousness experience is a perception of timelessness. Pure consciousness is experienced as beyond all forms and time, and is seen as equally present everywhere. Pure consciousness does not recognize separation. In the enlightened state is a "oneness where there is no division into parts," writes Hawkins. In pure consciousness there is the cessation of ordinary flow of thoughts and feelings. It is a state of infinite power, compassion, gentleness, and love.

Life is the expression of consciousness in the observable experiential world of form. An individual's level of consciousness is determined by the principle to which one is committed, writes Hawkins. To maintain progress in consciousness, there can be no wavering from principle. Once a principle is broken, it can propagate like cancer. For example, a society that supports killing, such as wars or death penalty, will not be able to stop criminal killing.

Highly conscious beings can affect society. Hawkins' kinesiology studies have shown that a single avatar at the level of

1,000 can totally counterbalance the collective negativity of all mankind. An individual at the level of 700 can counterbalance 70 million people below 200. One individual at level 400 can balance 400,000 individuals, and one at level 300 can counterbalance 90,000.

After completing his research, Hawkins has come up with some profound conclusions. He said there is no conceptual artifact between creator and created. If we step outside time there are no causes at all. The Big Bang can only occur in the mind of the observer. There is no beginning or end to a hologram, and we live in a holographic universe. It is everywhere and completed. There is no enfolded and unfolded universe, only a becoming awareness. The notion that there is a knower and a known is in itself dualistic, as it implies separation between subject and object. Existence is therefore merely a statement that awareness is aware of its awareness and of its expression as consciousness. There is only one absolute truth, all the rest originates from the artifacts of limited perception. There is ultimately neither duality nor nonduality. There is only awareness. Awareness is beyond consciousness. Those who have attained awareness report that it can't be described. This is the true statement of reality, universally and eternally, and is the essence of enlightenment.

Finally, Dr. Hawkins has provided a simple roadmap to achieve pure awareness.

1. An intense desire to reach awareness.
2. The discipline to act with constant and universal forgiveness, kindness, and gentleness, without exception.
3. Be compassionate toward everything, including one's own self and thoughts.
4. Be willing to hold desires in abeyance and surrender personal will at every moment.
5. Let go of the desire to own thoughts and silence the mind.

The level of consciousness in Dr. Hawkins' book *Power Vs. Force* rated was 810, suggesting the great truths that this remarkable man has uncovered on his quest to understand consciousness.

Dowsing

Dowsing is another method of accessing the implicate order of universal consciousness. Dowsers often use dowsing rods when searching for subsurface water or oil. When the dowsers recognize a movement or change in energy in the dowsing rods, the dowser identifies the location of water for a well. The dowser asks the question with his mind and the dowsing rod responds. Similar to the dowsing rod principle is the pendulum. The practitioner can ask a yes or no question with his or her mind. A "yes" response will cause the pendulum to swing in one direction or in a circle, and a "no" response will cause the pendulum to move in the opposite direction. As in applied kinesiology, any yes or no question will be answered by the pendulum. It can access the great data bank of knowledge of everything that has happened in the universe that lies awaiting in the implicate order or Mind Matrix. Often practitioners can dowse for a percentage by asking a "yes" or "no" question.

Dowsing is an ancient practice that is still used today to find subsurface water and oil. Standard Oil, Getty Oil, and Rothschild Oil have all used dowsers to assist in finding oil as well as high volume water sources for oil refineries. Municipalities such as Flagstaff, Arizona, have used dowsers to find water for the city.

A number of studies have been conducted on dowsing and water exploration. One study in Sri Lanka found a 96 percent success rate of 691 wells drilled, compared to a normal success rate of between 30 and 50 percent using traditional methods. Dowsers can also identify depth and flow of water. One study in the Congo found dowsers were able to predict the depth to within 10 percent accuracy in 67 percent of wells. German studies found that dowsing methods outperformed scientific approaches in 2,000 well drillings. The U.S. military used dowsing in Viet Nam to help locate and identify underground tunnels built by the Viet Cong. Using dowsing, one oil company had 80 percent success of 35 wells, compared to a normal success rate of four percent. Some water and oil dowsers never go to a site. They use a pendulum to dowse a map, from thousands of miles away. They can also determine the depth and flow by dowsing a map.

Research of the dowsers themselves found there are subtle movements in the muscle of the arms or hands which cause the dowsing rod or pendulum to move. It is a muscle movement that causes movement, not the force on the pendulum or dowsing rod. Researchers conclude that it is not an electromagnetic energy but a psychic energy that provides the information.

Changes in brain wave patterns occur in the dowsers, and messages are registered in the subconscious and parasympathetic nervous system. Dowsers always ask yes and no questions. Most times the answer comes back "no," but when the message comes back "yes," it causes a change in the parasympathetic nervous system causing a change in muscle tone and tightening amplified by the dowsing rod or pendulum. Change in muscle tension reflects change in the subconscious mind.

We all have the ability to be dowsers who can tap into the universal database of knowledge. Dowsers are able to tap into the torsion field of the implicate order that houses universal knowledge. Any type of target can be identified by dowsers. Both dowsing and kinesiology can access universal knowledge. They can answer questions regarding science, religion, consciousness, or any other question that can be answered "yes" or "no."

Radionics

Another method used to access the universal consciousness of the implicate order is radionic devices. Like dowsing, these devices can operate over thousands of miles with no decrease in accuracy. The methodology of the radionics requires an operator, a radionic device with dials to record frequency, the subject, and a "witness" that is a sample of tissue (such as blood or hair) from the subject. Often radionic devices have been used for health problems. In the early 1900s, studies by Dr. Albert Abrams of Stanford University determined that each disease has a signature vibratory component. It is thought that the radionic device is picking up this resonance.

The radionic mechanism is not well understood, but it appears to use subtle energy and resonance between the practitioner, subject, and device. Physicist Claude Swanson believes it may be the coupling of subatomic particle spins by way of the torsion field. T. Galen Hieronymous developed a radionic transmitter which actively generates energy that is reflected back from the environment. It is believed that a well-defined resonant frequency occurs that is recorded by the instrument. Hieronymous' research confirmed that cells emit resonant frequencies and that a subject's health and illness can be affected by interacting with these resonant frequencies.

George de la Warr of Oxford developed and researched many radionic devices. He also determined resonate frequencies of numerous substances that could be tested by the radionics device. Another researcher, Dr. Zaboj V. Harvalik of Lorton, Virginia, is a well known dowser researcher with a background in physics and is a former scientific adviser to the U.S. Army's Advanced Materials Concepts Agency. He concluded from his research that radionic devices are fancy dowsing devices. They depend on the physiologic response of the practitioners asking the questions. The witness (such as a hair sample) provides a resonant link to the subject. The "stick plate" or "rub plate" is used for detection purposes. The practitioner rubs a finger on the plate at various dial settings and notes what value the friction changes. The dials are used to set "rates" based on the Warr frequency tables. The vital element is that the mind of the operator must hold the question or pose the question during the process. Harvalik claims the elaborate technology of the radionic device helps the operator focus on the question being asked. The real detector is in the friction experienced by the practitioner's fingers on the stick plate. When there is noticeable change in friction on the fingers as they are being moved on the stick plate, the resonant frequency of the question being asked is recorded on the dial. It is another method to access the subconscious mind of the operator who can access psychic information from the database of universal knowledge. Radionic machines are used to send repatterning information with the intent of rebalancing the subtle energy field of the subject.

Evidence has been presented that there is is reservoir of knowledge of everything that has happened lying in the collective conscious. Pioneer researchers such as Dr. David

Hawkins and others have shown scientifically that it is possible to access this universal knowledge by posing yes and no questions. Kinesiology, dowsing, and radionic devices all appear to be accessing the data base of the Mind Matrix.

Chapter Eight

THE PARANORMAL WORLD

A New Science Explanation

Paranormal events have been noted forever, but have not been explained by traditional science. Science has divorced itself from the study of paranormal and spiritual phenomena. Perhaps it's time for traditional scientists to expand their box and try to explain paranormal happenings. They don't have to look very far because quantum physics is in their backyard. Of course, funding and preserving professional reputations are also factors. The science of quantum physics has been around for over a century and David Bohm's holographic theory for nearly a half-century. David Bohm was the first world-class scientist to suggest that a holographic universe might provide the science underlying the paranormal. We have already discussed the science underlying remote viewing, kinesiology, dowsing, psychokinesis, acupuncture, etc. These disciplines are emerging from the paranormal closet into traditional science.

There are other areas of the paranormal that can be explained by the new holographic science of Oneness. These include out-of-body experiences, near-death experiences, ghosts, channeling, telepathy, clairvoyance, teleportation, and the appearance of disincarnate entities. All have to do with consciousness, as does everything else in this world. If we are all interconnected, as quantum physics suggests, and if consciousness is eternal, as is also suggested, perhaps we can tap into the higher consciousness of entities that might provide the knowledge and wisdom to facilitate our spiritual journey in human bodies.

A number of researchers outside the confines of traditional science have been able to communicate with evolved, disincarnate beings that are helping us on our life's journey. Some also have been able to tap into the database of the Mind Matrix of higher consciousness. Edgar Cayce, one of the best known American clairvoyants, accessed this higher consciousness. From his many readings while in a trance state, volumes of material have been gathered from the information he received. The Association for Research and Enlightenment, located in Virginia Beach, Virginia, houses all of the Cayce readings about the past, present, and future probabilities of civilization.

Helen Schucman, Ph.D., professor of psychology at Columbia University, was able to communicate with Jesus, who dictated a book to her entitled *A Course in Miracles*. Jesus told Helen the same thing David Hawkins discovered in his research, e.g., we are not victims. One needs to just change attitudes and perceptions about difficult life situations that cause a sense of separation. The change in perception is a miracle.

Dr. Norma Milanovich, while a professor of education at the University of New Mexico, began to receive information from a highly-evolved extraterrestrial civilization called the Arcturians. They are fifth dimensional beings who have monitored the evolution of Earth's civilization. Eons ago they once inhabited this planet before going to Arcturus. The Arcturians were once third dimensional beings, like humans. Milanovich was told that when their civilization understood the concept of Oneness, these former Earthlings were able to evolve out of the third dimension into the fifth dimension. Third dimensional life was no longer a challenge. Milanovich was told this window of opportunity will again open around the time of 2012 for humans to evolve into the fifth dimension. In order to make this transition, humanity needs to understand the Law of One. The Arcturians are confident that we will become fifth dimensional beings. If we miss this opportunity, we will have to wait another 26,000 years for the next window of opportunity.

The Seth Material

Another person who received messages from a highly evolved being is Jane Roberts, who wrote a series of books about the information she received from a disincarnate entity named

Seth. Seth explained through Roberts the interdependence of consciousness and reality that corresponds closely to what has been written about so far in this book.

Seth told Roberts that humans have created their reality. We do not realize it because at a certain stage in every species' evolution this process is embedded so deeply in the unconscious that the individuals are completely unaware they are the creator. Seth refers to this condition as the "camouflage" system of reality. Seth said we inhabit a "matter concentrated" system of reality. As long as we remain ignorant of the fact that consciousness is the sole creator of the universe as we perceive it, we will remain seemingly entrapped in a plane of existence in which cause and effect seem to be our primary operants.

Seth explains that our own focus of consciousness is like a tuning system and that there are many reality pictures that we can tune into. Seth emphasizes that we need to fully understand that **thought is the ultimate substance of everything.** Physical objects are quite literally, like words or symbols, another medium of expression. He says we can create physical objects just as surely as we create words. Objects are natural by products of the evolution of the species, just as words are.

Seth states that humans in a natural state of evolution will began to perceive other levels of reality, levels in which pure information of thoughts reign supreme. In these levels of reality, cause and effect, as we perceive them, and time and space, as we know them, will no longer be of primary importance. Humans will discover that our unconscious minds are already connected to these levels. It is our conscious minds that edit out our awareness of them because our intellectual capabilities are not yet equipped to handle the volume of data encompassed by this interconnectedness.

Seth claims that every earthly event does play itself out in all probable states. Each quantum particle as well as every human being and historical event does have an infinite series of probabilistic counterparts. We are infinitely interconnected with these parallel universes, and we are multidimensional, as is Seth. The only difference is that the focus of our awareness is just not yet tuned into this fact.

It is impossible to separate one physical event from probable events, for they are all dimensions of one action. Seth said it is basically impossible to separate the "you" that you know from the "you" of which you are unaware. He said there are always inner pathways leading between probable events, because all of them are manifestations of an act in its becoming. He said reality is the one in which the perceiver is focused. All others would seem peripheral. Any given reality system will be surrounded by its probable clusters.

Each consciousness depends on every other. The strength of one adds to the strength of all, and the weakness of one weakens the whole. Also, Seth said, the energy of one recreates the whole, and the striving of one individual increases the potentiality of everything that is. This places a great responsibility upon each consciousness.

Seth emphasizes that we should change our bias that we are somehow separate from the fabric of the whole. Every being possesses a separate and individual focus of consciousness, but each of us is also part of a vast symbiosis, an infinite interpenetrating network, a dance within a dance within a dance. Seth explains that he was both an extension of Jane Roberts' constellation of consciousness as well as an individual focus of consciousness unto himself.

Seth said that consciousness is in all things, and thus we are different in degrees, but not in kind, from all those other self-organized aspects of life we recognize around us. There is no fundamental spiritual difference between humans and the "creaturehood" from which humans have evolved. Even in subatomic particles, there is consciousness, confirming quantum physicist David Bohm's hypothesis.

Seth explained that the degree of consciousness manifested at the subatomic level should not be thought of as processing "human characteristics." It would be better to view these constituents of matter as an inclination toward living things.

Seth confirms that consciousness is separate from the brain, but is deeply intertwined with it and controls it for the duration of its physical incarnation. Seth acknowledges that genes and chromosomes contain the information necessary for human existence. However, information is conveyed upon the DNA in another dimension entirely, impressed upon it at the time of material formation or incarnation.

Seth also said that consciousness has a quantum structure with its corresponding wave radiation. These quanta of consciousness are so small that they are beyond the range of physical matter as we know it, implying they lie beyond the quanta, as physicist Claude Swanson has suggested regarding torsion particles. The quanta of consciousness are already endowed with quanta uncertainty or the inherent probability allowing for infinite patterns and fulfillment in the reality system they compromise. **He emphasized that the quanta of consciousness comprises the substance of all beings and things.**

Seth commented on God, saying the concept is tricky. "The concept of God that you have has gone hand in hand with development of your consciousness. The ego, emerging, needed to feel its dominance and control, and so it imagined a dominant god apart from nature . . . The god concept was then an aid, and an important one to man's emerging ego."

He confessed to Roberts there is an organization of intelligence as far in advance to him as he is to us, suggesting he may not always be 100 percent correct. He is explaining to Roberts the knowledge that he has.

Author Norman Friedman integrated the Seth material with quantum physics in his book *Bridging Science and Spirit.* Seth views consciousness as the basis of all reality and he agrees with Carl Jung that the outer ego is not aware of the unconscious directly. Seth defines outer ego as our normal waking consciousness. He said the inner ego or inner self organizes the inner reality of the collective unconscious, thereby creating the substance we call the third dimensional world. The unconscious mind or inner ego is much more active and knowledgeable than the outer ego. The outer ego arises out of the inner ego. He said the inner self has access to a huge database of knowledge, the collective unconscious, and through its creation of the physical world provides stimuli to keep the outer ego constantly alert and aware. Seth provides a metaphor in which the inner ego writes a play and the outer ego acts it out.

The collective unconscious is very similar to Bohm's implicate order. The outer and inner ego together organize and manipulate the collective unconscious with its immense repository of knowledge to produce the third dimensional world. Bohm's model of the vast reservoir of knowledge in the implicate order (the inner ego) produces the third dimensional world – the explicate order (the outer ego).

Seth said the outer ego has available to it a more restricted information system in order for it to accomplish its tasks more efficiently. The inner ego has a broader context of the information because it has access to the collective unconscious or Mind Matrix.

Seth said the inner ego is a higher dimensional consciousness which cannot be contained in third dimensional reality. He said the number of dimensions the inner ego can reach is infinite, suggesting that itself is a projection of still higher consciousness. This higher consciousness is what Seth calls the soul, emphasizing the soul is not a closed system. Seth says there is a deeper identity who forms the inner ego, while the outer ego makes the decision about incarnation and in the place and time. The soul is the core of our identity from which we have sprung, the multidimensional personality of which we are a part. Seth claims there are no real divisions to the self, but speaks of various portions to help us understand the concept. The soul can more fully express itself and grow by experiencing lower dimensions.

In our system of reality, says Seth, we are learning what mental energy is and how to use it. We do it by constantly transforming our thoughts and emotions into physical form. We can get a clear picture of our development by perceiving the exterior environment. What seems to be a perception, an objective concrete event independent from ourselves, is instead the materialization of our own inner emotional energy and mental environment. In other words, the objects that we perceive are actually created by our inner ego and are our own symbols through which we express emotions and thoughts. This is how we develop consciousness, explains Seth.

Seth said that our thoughts and emotions are formed into

physical matter by definite methods and through specific laws not presently known to us. To help us understand matter, Seth describes basic building block units of consciousness he calls consciousness units or CUs. They are not physical, but can be in all places at all times simultaneously. These CUs contain within themselves innate properties that can expand, develop, and organize, but they always maintain a kernel of their own individualities. These CUs don't have human characteristics, but have their own learning inclination. They are unpredictable, allowing for infinite patterns and fulfillment. Seth said CUs are divine fragments of ALL THAT IS. Consciousness, said Seth, is the basic constituent of the universe. It is the very essence of the world. Seth said that whatever organization CUs become part of, or how they mix with other basic CUs, **it is our identity and is not annihilated.**

Seth explains that CUs can operate as either particles or as waves, confirming Bohm's theory – in the explicate order they are particles and in the implicate order they are waves. Seth refers to the psychic energy of ALL THAT IS, with CUs being part of the energy, similar to Bohm's holomovement.

Pure energy is not divided. It contains within it the propelling force toward the formation of all possible variations of itself. Seth said the smallest unit of pure energy contains within itself no mass, but has the propensity for the creation of matter in all its forms with the impetus to create all possible universes. This is the Mind Matrix, zero point field, implicate order, chi, etc., that we have discussed. Bohm said the wholeness in the holomovement cannot be divided. CUs can take on form and become separate entities (ripple), or they can flow together in a vast harmonious wave (ocean) of activity as a force.

Seth said that each "particlized" conscious unit rides the continual thrust set up by the fields of consciousness in which waves and particles belong. Each particlized CU inherently contains within it the knowledge of all other such particles. At other levels, the CUs operate as waves. Again, Seth's statements are supported by quantum theory. The wave and particles are but one, the particle being a localized manifestation of the quantum field. The particle contains within it the knowledge of all other such particles, which is the hologram. CUs are the building blocks of our reality.

Seth said there are universes that exist alongside us, outside our reality. At some points, different realities do coincide. Seth calls them coordinate points, which act as channels through which energy flows. They act as warps or invisible paths from one reality to another. These coordinate points represent accumulations of pure energy. They are activated by thoughts and emotions. The intersection of thought and coordinate points makes possible the projection of the thought into physical matter. Seth said the intensity of the thought matters, not the nature of the thought. Matter is formed through the relationship of CUs and coordinate points. A combination of CUs forms a consciousness.

The subjective experience of consciousness automatically expresses itself as electromagnetic units or EUs created by subjective experiences, such as thought, mental images, and emotions. A strong emotional component makes it more likely that mental activity will be manifested in physical form. Electromagnetic energy units originate from consciousness and exist just below the range of physical matter. Electromagnetic units activate the dormant energy of a coordinate point and manifest into matter. Seth explains that consciousness is shaping consciousness, and matter is a form of consciousness.

What Seth describes confirms what Bohm believes, that "Mind and matter are inseparable, in the sense that everything is permeated with meaning . . . ." At no stage are mind and matter ever separated. They are different levels of mind.

Seth explains that electromagnetic units operate beneath the surface of our reality and can travel faster than the speed of light. When they do become matter, they are slowed down to the speed of light and appear in solid form. These electromagnetic units originate as mental activity and are the building blocks in matter. In other words, mental activity provides the blueprint for the object. Electromagnetic units arise from all levels of consciousness, not just human consciousness. Seth's idea that mental activity is required to activate the coordinate points is similar to Bohm's implicate order, in that mind provides the stimulus to form the field of matter. Bohm said the quantum wave carries information and is, therefore, potentially everywhere. It is only active when this energy enters into the energy of the particle.

Probability Realities

Seth said that all actions are initially mental actions and all mental actions are valid. In daily life, we have many choices of actions which actualize into our reality. The actions we do not choose are as valid as the ones we do choose. The unchosen action actualizes elsewhere in other probability realities. This means that in other realities, there are other probability selves using actions that we have not chosen to use in this reality. Not only do our thoughts produce worlds, but each of these new worlds is producing worlds of its own. Seth said that probabilities are an ever present portion of our psychological environment and there are profound interconnections binding the probabilities to each other. The connection occurs beneath normal consciousness through a telepathic framework. The inner ego is what determines the information coming into our consciousness. According to Seth, communications take place between various probable selves in dreams.

Seth said our greater identity is in touch and aware of all probable selves. Our uniqueness in the universe is very important. We seek out and view experiences with our own special affinities.

In the Beginning

Seth explained that in the beginning there was no God. In fact, there is no beginning. Our world was created from inside out because consciousness was before matter. Consciousness materializes as matter in physical life. Seth said our own consciousness and our collective consciousness are part of the divine structure. Every part of the universe is conscious.

Bohm sees the beginning of the universe as does Seth, except that it is in a timed event. Bohm's universe is a little ripple arising from a vast ocean of energy, the implicate order. Seth said that our visible universe results from electromagnetic units from the invisible universe and becomes manifest. Bohm sees the Big Bang as a ripple while Seth sees the Big Bang as manifesting in our minds. The director of Bohm's creation is the implicate order, for Seth it is the inner ego.

Seth said that ALL THAT IS, the Absolute, the holomovement in Bohm's terms, is a constant presence that surrounds us. He calls both religion and science a myth. A myth when accepted by culture is considered fact. Seth said our future personalities are as real as our past ones. We are all interconnected manifestations of the whole, but at the same time every one of us is but a unique materialization like no other.

Sai Baba

It is said that those who inhabit the fifth dimension are able to manifest anything with their minds instantaneously. Humanity will eventually evolve to this dimension. It is important that we monitor our thoughts because we do create our reality, but in the third dimension it takes longer to manifest. In an ashram in southern India resided a very evolved spiritual leader named Sathya Sai Baba (Sai Baba passed away in April of 2011). Sai Baba had the ability to manifest material objects from empty space or from the Mind Matrix. Sai Baba could manifest lockets, rings, jewelry, and watches out of thin air, which he then distributed as gifts. He could also materialize an endless supply of Indian delicacies and sweets. From his hands, Sai Baba could manifest sacred ash called *vibuti*. Many followers of Sai Baba have pictures of him, and frequently *vibuti* will manifest on the picture thousands of miles away.

Thousands of individuals, including scientists, have observed Sai Baba materialize objects. When co-founder of the International Association for New Science, Dr. Brian O'Leary, a former Princeton astrophysics professor and Apollo astronaut, visited Sai Baba, he manifested a watch for Brian out of empty space. Neither magicians nor scientists can explain how Sai Baba does this. Psychologist Erlander Haraldson from the University of Iceland had spent over a decade studying Sai Baba and published his findings in a book entitled *Modern Miracles: An Investigative Report on Psychic Phenomenon Associated with Sathya Sai Baba.*

Sai Baba can materialize specific objects on request. He has materialized large quantities of food, with some foods so hot that people can't hold them. These miracles are similar to Jesus' feeding the masses. As Sai Baba walked among his followers, great amounts of sacred ash are manifested from his hands. Over the many decades that Sai Baba had been on the planet, no one has found evidence that Sai Baba has faked his abilities.

Sai Baba also had the ability to bilocate. Witnesses have seen Sai Baba snap his fingers and vanish. Instantly, he reappears one hundred or more yards away. Such evidence suggests that our bodies are not objects but holographic projections. If scientists Stanislov Grof and William Tiller are correct and the mind is able to intercede in the implicate order, then anything is possible. The implicate order is the holographic plate that gives birth to the hologram we call the universe. The mind can create any reality of laws of physics it wants in order to make such things possible as demonstrated by Sai Baba. Tiller, who was head of the Department of Material Sciences at Stanford University, said the universe is an enormous "holodeck" similar to the "holodeck" on the televison show "Star Trek: The Next Generation."

Clairvoyance

The cosmic hologram of the implicate order contains information of all past events that have occurred in the universe. The record of the past is not stored at any one location. There are some gifted individuals who tap into this knowledge held by the implicate order or Mind Matrix. Clairvoyant Edgar Cayce, referred to as the "Sleeping Prophet" in many books, was able to enter a trance and tune into the past. An aide would ask Cayce specific questions while he was in trance, and Cayce would give a "reading" about the specific question. Upon awakening, Cayce could not remember anything from the trance. For example, he described the historical role of the Essene community at Qumran eleven years before the discovery of the Dead Sea Scrolls, which confirmed what Cayce had said. Psychics are able to access the past of many archaeological sites. The record of the past is stored in the Mind Matrix.

The future can also be accessed in the cosmic hologram.

Robert Jahn and Brenda Dunne demonstrated this at Princeton University. In 334 formal trials of the random event generator, volunteers were able to come up with accurate precognitive information 62 percent of the time. Even normal individuals who weren't considered psychic did well in the precognition viewing tests, according to Jahn and Dunne.

The holographic universe helps explain the ability of psychics. Bohm claims that every human consciousness has its source in the implicate order, implying that we all possess the ability to access the future. Psychics often report that precognition information often appears to them in three-dimensional shape.

Studies have shown that precognition occurs more frequently than any other psychic experience. Unhappy events outnumber the happy ones by a ratio of four to one with precognition of death leading the way, followed by accidents and then illness. Primitive cultures usually perform better on ESP tests than so-called civilized cultures. Between 60 and 80 percent of precognition occurs during dreaming.

Precognition of the future is not fixed in a hologram. People have the free will to change the event to avoid disaster if given a glimpse into the future. Some believe they may have slipped into a parallel reality. There have been nineteen documented cases of people who had precognition of the sinking of the Titanic. Some passengers paid attention to their glimpses into the future and survived.

Ghosts

Ghostly apparitions appear to be but holograms. In fact, not all ghosts are human. One anecdote involves a British military officer and his family watching an apparition of a horse-drawn carriage that pulled onto their lawn and stopped. Studies have shown that in the United States and England, between 10 and 17 percent of the general population have seen an apparition.

Many hauntings have occurred at locations where some act of violence or powerful emotional event had taken place. In addition to images and sounds, the emotion being felt during the occurrence is also recorded in the cosmic hologram. Emotional intensity of the event makes a more pronounced imprint on the holographic record.

Near-Death Experience

Both near-death experiences (NDEs) and out-of-body experiences (OBEs) appear to be holographic phenomenon, writes Michael Talbot. When a person has an OBE, he or she may not be traveling at all. Talbot thinks the "traveler" might merely be altering the always illusory hologram of reality so that they have the experience of traveling somewhere. In a holographic universe, not only is consciousness already everywhere, it also is nowhere, says Talbot. OBEs are more common than one thinks. Dr. Fowler Jones, a psychologist at the University of Kansas, randomly selected 420 subjects, and 339 of them reported they had experienced an OBE. Jones concluded that OBEs are expressed by normal, healthy, and intelligent people. Stanford Research Institute scientists Hal Puthoff and Russell Targ both acknowledge that OBEs are similar to remote viewing.

As mentioned earlier, Dr. Ken Ring, the world's leading researcher in near-death experiences, believes that near-death experiences can be explained by the holographic model. He thinks that NDEs and death itself are nothing more than shifting a person's conscious from one level of the hologram of reality to another.

Dr. Ken Wilbur said, "If the paranormal exists, it can only be understood through references to the implicate order, and there is no intrinsic reason why the paranormal should be impossible." Once we understand the concepts of quantum mechanics, the holographic model principles, and the Mind Matrix, paranormal events are normal and obey the laws of science. Closely related to the paranormal is the spiritual world, and as we shall see, the spiritual world can also be explained by the new science.

PART TWO

Revelations of Oneness

Chapter Nine

THE SCIENCE OF THE MIND

Oneness of Man with God

During the early 1920s, three spiritual visionaries received Divine Knowledge about the sacred attributes of mind and Oneness. They were Ernest Holmes (1887-1960), Charles Fillmore (1854-1948), and Walter Russell (1871-1963). Their revelations can be explained by the science of quantum physics that we have discussed in Part One. The information Holmes received from Divine Source was similar to Walter Russell's. Much of this chapter is based on Holmes' writings, especially from his book *The Science of Mind*. Holmes' information was so full of Truth that he felt his purpose in life was to share his knowledge through writings and the establishment of the Church of Religious Science, whose churches are found around the world, now called the Centers for Spiritual Living. Many people refer to the church as Science of Mind. Where did Holmes' information come from? For those who are familiar with Abraham and Esther Hicks, Esther was told by Abraham that Holmes' information came from Abraham, the group of disincarnate beings that have been providing sacred information to Esther for nearly two decades. The terminology is somewhat different from Walter Russells' but the information is much the same.

The Universal Mind

Holmes was told the mind of man is a part of the Mind of God. Within man's mind lies unlimited possibility of expansion and self-expression. The Mind of God is Infinite and man's mind is part of this Infinite, Creative Mind of God. Man's mind is as infinite as his capacity to understand his true relationship to God or Spirit.

Holmes' source said that Spirit is really the only Mind there is. The Universal Mind is eternal as it had no beginning nor ending. Spirit is the only Conscious Intelligence in the universe and is the only Directive Intelligence in the universe. Because the mind of man is the Mind of Spirit (God), man's mind is conscious and directive. Man's mind is forever evolving into a greater awareness of its role in the creative order of the cosmos.

Holmes was instructed, as was Walter Russell, that God is Mind and God is Spirit. The self-knowing Mind of God is the Spirit of God as well as the Spirit of man. Spirit is Mind in its self-knowing state. The conscious mind of man is part of the self-knowing Spirit that operates through the thought of man. This conscious mind of man is that part of Unity with Spirit (God). It enables man to be an individual entity who has a separate identity without being separated from the Whole Spirit (Mind) of God. Without this individualized conscious mind of man, God would have no independent offspring through which He could be expressed. God has placed Himself in man's being in order that man may function individually. This Divine self-knowingness in man enables man to work out his own destiny and at he same time to be a way for God's expression.

Holmes wrote that "the subjective mind of man is part of the Universal Subjective Mind of God." This allows every human to individualize himself or herself. The Universal Mind also falls under a law called the Law of Mind. This Law includes the Law of Cause and Effect and the Law of Karma. This Law of the Whole is now individualized by man. In man's earthly sojourn, the law may produce either freedom or bondage. Man's use of this universal Law often gives the impression that man's subjective mind is its own entity, not one with God. However, man's mind is always connected to the Universal Mind. It is a property of the individual in the Universal Mind. In other words, God and the individual man are One, but God is greater than man because the whole is always greater than its parts.

Holmes was informed that the Universal Subjective Mind is

"Mind in an abstract and formless state." This is Walter Russell's Mind Substance and Bohm's Implicate Order and Max Planck's Mind Matrix. It is the quantum field. It is a potential energy with an unformed latent power. However, it is always ready to take form. It is Mind and substance in an uncreated state of being ready to be molded into any form. It is unexpressed Creativeness, Power, and Substance. In other words, it is unexpressed Mind. Russell's metaphor for it would be a sea of Mind Substance. In its original state, it is invisible, but it has the potential for all possible forms (ripples in the sea). It is a creative universal energy waiting to be used and ready to take form.

Mind in its subjective state is Universal. It cannot act until it is set in motion by a conscious state. The Universal Subjective Mind knows how to create form but does not have conscious knowledge to do it or direct it. It is man's consciousness that calls it forth into expression.

Within the Universal Mind lies the essence of everything that ever was or will be. Both the visible and invisible are in it and governed by it. It is the only creative agent in the universe and any other apparent creative agents are the Universal Mind working in different ways.

Things exist in the Universal Mind as ideas which take form and become things in the visible world. It is our thoughts that call into expression the manifestation of things from the Universal Mind Substance. Holmes writes that Universal Mind in its unformed state can be summoned for individual use. This is the key to all proper mental and spiritual work from a practical point of view. Within the Universal Mind lies the potential for all our desires that lie in the unformed state. The formless is ready to take form. Because it is not self-knowing, it has no intention of its own and form must be imposed upon it. The self-knowing mind of man in the smaller world and God in the Collective Mind of the Greater World will create form.

The Universal Mind is an unformed substance and energy. It is energy plus intelligence. Intelligence is conscious energy that works upon the Mind Substance in accordance with Law. Man's mind can make a demand upon himself or upon the universe which flows through the self. It makes the demand on Universal

Mind and Universal Energy to produce certain specific things for him. He provides a creative force that produces all things in the universe. The mind that man uses to think new ideas is the original Mind of God. There are not two minds, i.e., one of man and one of God, but only One in Essence.

No form is present until humans think it into form. Forms are returned into the formless Universal Mind. All form is temporary and therefore an illusion. The formless is Eternal. Man has the ability to call these forms into being through his mind. In other words, we make use of the Mind of God through the Laws of the Universe. The conscious part of mind is limited but the subjective part of man is universal and without limit. The Universal Laws say that if we use our minds for destructive purposes, we shall cause destruction upon ourselves. There is no liability to use creative power constructively. Holmes wrote, "The Universe is foolproof." By tapping into Universal Mind through our thoughts, we can control conditions of our life, including healing ourselves. Both our body and events in life result from Mind held in form.

The mind of man is the Mind of God functioning at the level of man's understanding of his place in the universe. Holmes said, "The Highest God and the innermost God is One God." Through mind, humans can unify with the universe and utilize the Power in Universal Mind that can do anything for them. Humans have access to the original creative force of the universe.

Holmes' source said, "Spirit has already ordained that nature shall be perfect." The normal functions of life are harmonious but through ignorance man interferes with this harmony that results in discord, which is temporary. This is where the Law of Cause and Effect comes into play. We are responsible for any discord that we encounter. It is ignorance that causes self-imposed bondage of man, but it is knowledge that can set man free. Humans have been given free will. The Universal Mind, as Law, knows about the individual only when humans know about themselves. It is the desire of humans that guides the Universal Mind. Desire is what decrees Universal Mind to manifest for the individual.

Thought is the energy that forms the universe into creation. The Law of Creation depends on the power of human desire. The desire of thoughts mold our ideas and beliefs into visible forms.

The Law Of Mind, Thought, and Causation

The Universal Mind has unlimited power. For humans, Universal Mind has only the power that humans give to it. Universal Mind is a force in nature that is to be used and brought under conscious control by the Spirit of man in his conscious mind. Man's conscious mind is the Consciousness of God within him. Universal Mind is helpless without direction and must be directed or it will be useless. Holmes writes that "Man is the consciousness of Spirit individualized, the personality of God multiplied." In other words, Spiritual man is One with Spiritual God. Spirit works through man as self-knowing, making humans different from other creatures.

Universal Mind is unconscious Intelligence without desire that lives only through man. Man is the desire, will, and purpose. Universal Mind has Infinite Intelligence within itself. Put in other words, it is Mind in an unthought state. From desire and mental images it molds form that is impressed upon it. Universal Mind will do whatever we desire of it, but first we must create the thought mold. The Law of Mind obeys the thoughts given it, whether we are conscious or unconscious. Universal Mind must obey conscious thought because it has no power other than deduction. Conscious thought has so much power, and this part of man's mind is part of God's. Man's mind can be considered as God being individualized.

The Universal Mind is compelled to receive the image of man's thought. Essentially, man's thought will direct Universal Mind. Whatever exists as a mental picture in Universal Mind must eventually take form if it is really believed in and embodied. This is why it is so important to think clearly to allow the image of our thoughts to sink into the subjective state of Universal Mind. Conscious thought is the starting point of all new creations. Holmes emphasizes that trained thought is much more powerful than untrained thought. The subjective Law of

Universal Mind knows only to obey. Holmes writes about tools for proper use of thought which include:

> "We do not wish, we know.
> We do not dream, we accept.
> We do not pray, we announce.
> We do not expect something to happen, we believe it has already happened."

When we use our subjective mind, we are using it as a creative power of the universe. There is only One Mind, not two; the individual mind is part of the Universal Mind.

Holmes wrote that mind in its conscious state is Spirit, whether we think of it as man or God. Mind in its unconscious state or subjective state is the Law of Cause and Effect. Any effect can be changed by altering our thought relationship to the fundamental Law which gives rise to lesser causes and effects.

The Law of Mind knows how to make things out of ideas. Holmes emphasized that when humans think about lack, poverty, and limitations, these thoughts provide the conditions to create lack, poverty, and limitations. Likewise, thoughts of abundance will manifest as success and happiness. The Law of Mind is not selective; it is receptive without caring what it receives. It is also creative without caring what it creates. Conscious Mind or Spirit controls the Law of Mind or Universal Mind. One can reverse the thoughts and the present conditions will tend to reverse.

Ideas are nothing but form, conditions, circumstances, cause and effect, and everything between cause and effect, wrote Holmes. Fear is the great enemy of man, which is the reverse attitude of faith. Perfect love will eliminate fear. It is ignorance that has caused us to experience fear. The Laws of Mind come under control of the Spirit; with the will of Spirit comes peace, clear thinking, and happiness. The Spirit guides, and the Law executes. Spirit is present in its entirety at all times and all places. Pure Spirit exists at the center of all form. Wherever we center our attention, its image must concentrate the Mind Substance into the form of our attention. The Universal Mind is formless but it always gives birth to form. The form comes and goes but the Universal Mind and Spirit go on forever. We are part of it, as there is no limit to Spirit's manifestation except for limitations we set. We are all connected to the Law of Universal Mind and therefore connected to the creativity of the Whole.

Mind is the realm of all causes, and conditions are in the realm of all effects. The instrument of Mind is thought, and new thoughts create new conditions. Spirit or the Mind of God is all causation. Spirit or Mind is also all effect. All form is temporary, even though it can last for millions of years. Cause and effect are considered the two ends of unity. The physical end is visible, while the spiritual end is invisible. Holmes wrote, "The ideas of God are perfect laws on their way to producing complete and perfect effects." Mind is always conscious, with its consciousness being its Law. Both consciousness and Law are perfect. Mind can only be conscious of itself, because it is All.

Generic man is held in the Mind of God as a complete and perfect manifestation of the Divine. Mind is all inclusive and pervades everything. It is both self-governing and self-propelling. Love and Law are a tandem force that produces a complete and perfect manifestation. It is man's consciousness of God that constitutes his real and immortal self.

Holmes' Divine Source said the relationship between the individual and Universal Mind is one of reflection. The laws that have created bondage can just as easily be used to create freedom. Humans can give attention to any one idea which will automatically be embodied. We will attract to ourselves the objective likeness of this embodiment and therefore the thought becomes a thing. The objective form is nothing but the reflection of our subjective thought.

Holmes emphasized the lesson of the Law of Cause and Effect is simply that the Universe is One. It is never two. It is foolproof and never operates against itself. If desires are destructive, one shall suffer from them until the lesson is learned.

Spirit is never antagonistic toward mankind as it is always expressing itself through us. Because man and God are One, the self-expression of man is the self-expression of God. The "I Am" principle is both individual and universal. All form is rooted in the Creative Mind, and the Spirit of God is but the Spirit of man.

Holmes wrote, "To practice the Presence of God is to awaken within us the Christ Consciousness. Christ is God in the soul of man." He says the resurrection is the death of the belief that we are separated from God. Death is an illusion and not reality as man awakens to Life.

Thoughts are more than things as they are also the cause of things. There are no material causes or effects. All things are but illusions. They have no independent existence since there can be nothing external to a comprehensive mind. All our work is done in the Mind only, as our entire vehicle for work is thought and the knowledge of the power it utilizes. This power is much greater than the intellect in its creativity.

Oneness

The unifying principle of life is that we all exist as a One all-embracing Mind. Through it we all live and subsist. It is the Mind through which we live that is independent of any individual on our part. A Divine Partnership exists between the seen and unseen. Our real self is God (Higher Self) connected to the Universal Mind, and through the center of God consciousness good will come to us.

A Universal Intelligence coordinates everything into one complete Unity. The physical universe is always mechanical, whereas the Spirit is always spontaneous. Spirit is Unity and can never do anything that contradicts its own nature. There is a universal standard of Reality that we all will discover someday. The Law of Mind is so simple. If we think abundance, we create abundance and vice versa.

Creation is the result of man's sojourn in life and self-expression. The One encompasses everything and flows through all, resulting in countless forms and personalities. The Divine, of which we are a part, will bring all creation into a state of perfect manifestation. We will do this through cosmic consciousness, meaning one's consciousness uniting with the Whole.

As we become conscious of God and Oneness, evil will disappear. Mystics have always said there is no ultimate reality to evil. It is only an illusion. There is no devil; there is no damnation except for one's state of thinking; there is no punishment outside of self-infliction through ignorance; and there is no salvation outside of conscious cooperation with the Infinite.

Evolution is an effect of intelligence and not its cause. Evolution always follows involution. Illumination will come to mankind as humanity realizes its Unity with the Whole, allowing the Truth to operate through it. Jesus taught that as we consciously enter the One, the One becomes us. The individuality of man is a Divine Reality. All the mystics have known about the absolute Unity of God and man. Within the Universal Mind each individual exists, not as a separate entity but an individualized center of God consciousness. The Spirit of man needs no healing because the Spirit of man is God. Every person's stream of consciousness is connected to the One Universal Mind, and anyone tapping into this stream of consciousness may objectify it.

Everything that has ever been thought, said, or done is retained in the field of unity in the Universal Mind. Time and space do not exist within Universal Mind. The past and present are One. Everyone who has ever lived has imprinted a mental picture of themselves within the Universal Mind. We are all One with the Father, and we all represent Him in an individualized state. So ends the synopsis of Ernest Holmes' knowledge of the Universal Mind. He was a very wise man. As we have seen, science has substantiated the writings of Ernest Holmes.

UNITY

Another spiritual visionary who taught the concept of oneness was Charles Fillmore, who lived in the same era as Ernest Holmes and Walter Russell. His divine revelation was similar to that of Holmes and Russell, that we are all connected. Fillmore wrote that "All things, all actions, all principles are working toward the unity of God, man, and the universe." It was his strong belief in the unity of all things that motivated Fillmore to found the Unity Church, headquartered in Unity Village, Missouri. This section is based upon his writings.

Fillmore wrote that God is Mind, and man is made in the image and likeness of God. The One Mind of God is the only creator. The One Mind is the Source, which is the source of all reality and originates all that is permanent.

As revealed to Holmes and Russell, Fillmore's source said that Mind is the reservoir of the universe. It provides all that appears in the universe. From this reservoir of Mind, "We draw all that we make up into our world, and it is through the Law of Mind that we form our lives." Fillmore emphasized that it is important to understand the Laws of Mind because man is the expression of Mind. Man has the ability to consciously draw divine ideas from the One Mind. Fillmore writes that all wisdom is implanted in us by Divine Intelligence, which is another name for God. "The one supreme Law of God," wrote Fillmore, "is spirit manifesting itself in the mind of man." It is necessary that we understand our own importance as being a manifestation of God. We are not separate nor insignificant but a vital, important, and integral part of a mighty whole, wrote Fillmore.

Fillmore tried to bring clarity to the Bible and Jesus with the principles of Oneness. Material things are condensed thoughts, and matter contains a vast amount of hidden energy. Jesus understood this principle, allowing him to break the bonds of atoms with his miracles by changing a few loaves of bread and fishes into many. Jesus recognized Mind is in everything and called it "Father." Jesus was able to be consciously One with the absolute principle of Being. There was no conscious separation of Jesus from the Mind of God, or Father, or Source, which are all the same. This super consciousness that Jesus externalized is omnipresent and is ready to manifest itself. We all have the potential to manifest this super consciousness. Jesus quoted Psalm 82, which said, "I said, 'Ye are gods and all of you sons of the Most High.'" Jesus taught that man is the light of God. God (Mind) is spirit and the omnipotent, omnipresent essence from which all things manifest. Mind is the source of all creation. Fillmore wrote, "There must be a unity between the mind of man and Divine Mind so that ideas and thoughts work constructively unto eternal life. . . ."

Thoughts

The creation of the One Mind is ideas. The ideas of God are potential forces waiting to be set in motion through thoughts of man. It is through the thinking mind that the visible universe has existence. Thoughts of man make the formless (Universal Mind) into form. It is through conscious cooperation with man that the One Mind forms its universe. Thinking is a process in mind allowing abstracts to be made concrete. The thinking faculty of man's mind represents man in the Godhead. Man is a co-creator with God through his thoughts.

Man is capable of seeing mental sights of ideas in Divine Mind and can proceed to put the ideas into terms of comprehension at his level of consciousness. All ideas have their origin in Divine Mind, but their development depends on man's knowledge of the Divine Mind. Divine Mind is the storehouse of ideas and man draws all his ideas from this omnipresent storehouse. Their form in consciousness depends entirely upon the individual's level from which he draws his mental image. Fillmore said that man cannot acceptably serve God or do His will until man understands fundamental thinking on how thought manifests into form. Man incarnates to do specific work in the physical realm and his first step is to realize his purpose. In other words, man is the instrument of God, a co-creator, through which he brings his potentialities into visibility. It is very important that humanity knows how we have produced this state of existence that we call life. Man should rapidly conform to methods that bring harmony and success into his life. We develop this harmony from communion with the God-Mind, a phenomenon Christianity calls the Holy Spirit.

The Holy Spirit in Divine Mind corresponds to our thoughts in our Minds. Fillmore said "The Holy Spirit means the Spirit of God." Remember, God is Mind. God's idea of His creation is His Son's, and this Son (idea) executes the plan of God (the Original Mind) which corresponds to our thinking in devising plans to execute the idea. When Divine Mind is brought into our world by our thinking, it is limited by our conception of it. Once we understand this relationship, we can approach God (Divine Mind) with confidence that we have access to these real, unlimited ideas. By cooperating with Holy Spirit, results manifest. Fillmore was told that "Christ is the Spiritual mind in every individual, and the Spiritual Mind is the offspring of the Universal Mind." Jesus said, "I and the Father are One." We all have the Christ Spirit within us and therefore we are also One with the Father.

Fillmore wrote that the subconscious mind is the realm that contains all past thoughts. The subconscious mind cannot take the initiative but depends on the conscious mind for direction. The mind when put in conscious contact with the Universal Christ Mind can release the energy stored up in the subconscious mind. We can all tap into this Christ Mind (Universal Mind) substance of which we are part. When we make contact with the Universal Mind Substance and it is released, this invisible current has unlimited potential. Man can overcome sin, sickness, old age, and all human limitations through thought. The Universal Christ Mind releases the energy stored up in the subconscious mind to create wellness and harmony.

The Body

The body demonstrates the different states of mind of those souls inhabiting them. It is the expression of one's mind. The body is the specific interpretation and expression of oneself. The Bible says that man is made in the "image" and "likeness" of God. The New Testament says that man's body is the dwelling place of the Spirit of God. The body is man's visible record of his thoughts. It is the individual interpretation of one's identity. An individual can become any type of being that he chooses. As a man thinks in his mind, so becomes his body. "The body executes the mandate of mind," wrote Fillmore.

Divine Mind has placed in the mind of man an image of the perfect-man body. If one sees the body as spirit-mind which is pure and incorruptible, one can prevent disease and death. Thoughts must be perfect to manifest perfection in the body. The body is the record of one's mind. Fillmore wrote, "The chief object of man is to glorify God in his body." There are no incurable diseases because "With God (Mind) all things are possible." All diseases can be healed by mind. Every idea is but a seed. It will manifest by the kind of mind soil in which it is planted. If one's mind has thoughts of sickness, one will manifest sickness. If thoughts of perfect health occur, perfect health will be manifested.

The real objective of our existence is to bring forth perfect man. Man must become conscious of the Spirit in God-Mind. This consciousness is soul and the tangible part of our soul. Jesus taught that we must attain the consciousness of eternal life. We can transform ourselves through thought, and only through thought will we realize the unity of the universe and our eternal being.

Summary

1. Man is a part of the Infinite Creative Mind of God.
2. Man is connected to the eternal Universal Mind.
3. Spirit is the only conscious and directive Intelligence in the Universe.
4. The mind of man is the Mind of Spirit.
5. Spirit is Mind in its self-knowing state.
6. Mind of man is part of unity with Spirit.
7. The Law of Mind includes the Law of Cause and Effect and Law of Karma.
8. Universal Mind is always ready to take form.
9. Universal Mind is set in motion by a conscious state of mind.
10. Man's thoughts manifest things from Universal Mind.
11. Within the Universal Mind is potential for all our desires.
12. The intelligent Universal Mind is an unformed substance and energy.
13. Man's mind provides a creative force to produce all things in the universe.
14. Form returns to the formless Universal Mind and therefore is an illusion.
15. If we use our minds for destruction, we shall destroy ourselves.
16. Man's mind controls all life conditions and healing.

17. Desire decrees Universal Mind to manifest.
18. Thought is the energy that forms the universe into creation.
19. Man's thinking mind is the Mind of God.
20. Man's conscious mind is the Consciousness of God within him.
21. Universal Mind is conscious intelligence without desire.
22. Man is the desire, will, and purpose.
23. The Law of Mind obeys the thoughts given it.
24. Man's mind is considered as God individualized.
25. The individual mind is part of the Universal Mind.
26. Mind in its unconscious state is the Law of Cause and Effect.
27. The will of Spirit is peace, clear thinking, and happiness.
28. Pure Spirit exists at the center of all form.
29. Thoughts are the cause of all things.
30. Mind is all inclusive and pervades all things.
31. Man and God are One.
32. We all exist as a one all-embracing Mind.
33. Universal Intelligence coordinates everything into one complete unity.
34. As we become conscious of God and Oneness, evil will disappear.
35. Within the Universal Mind, each individual exists as an individualized center of God Consciousness.
36. Man is the expression of Mind.
37. Thoughts of man make the formless Universal Mind into form.
38. The subconscious mind contains all past thoughts.
39. Through thought, man can overcome all disease, old age, and human limitations.
40. The body is man's visible record of his thoughts.
41. There are no incurable diseases because with Mind all things are possible.

Chapter Ten

MIND AND REALITY

Mind is the Whole Universe

The next three chapters are based on Walter Russell's book *The Universal One,* and the last chapter gives mundane meaning to Oneness. The information which was revealed to Russell in 1921, and published in 1926, is like a graduate course in unity consciousness. If you can grasp the concept of Oneness, you will have discovered the secret of the universe, life, the soul, and God. Some of it is difficult to understand, but I feel it is important to include it in the book. I apologize if I have not made it as clear and simple as possible, and you may have to read it several times to fully understand it, as I have used much of Russell's original language. Hopefully, the previous chapters have given you sufficient background to help comprehend it. Most of Russell's scientific revelations can be explained by quantum physics. Often I have repeated myself to help get the point across. Walter Russell was one of the first people to bring the science of Oneness to the world. Good luck!

Walter Russell

Walter Cronkite, former CBS news anchor, referred to Walter Russell as the "modern day Leonardo da Vinci." Born in Boston on May 19, 1871, Russell became a world class artist and sculptor. He painted portraits of many notable people, including Mrs. Theodore Roosevelt. His reputation as a sculptor was unsurpassed, through sculptures of Theodore Roosevelt, Franklin Roosevelt, Thomas Edison, and Mark Twain to name just a few. Russell also perfected his architectural talents by designing a number of New York City landmarks. His art studio was located in Carneige Hall in New York City. Some have considered Russell one of a handful of true geniuses who have walked this Earth.

Beginning at the age of five, for nearly ten years every May, Russell wandered the woods for up to ten days seeking the ecstatic joy of the soul. His consciousness stepped out of and into his body. Russell suffered a near-death experience in early adolescence after contracting black diphtheria, a condition which shut off his breathing. He was pronounced dead, and the undertaker arrived to claim his body. During the near-death experience, Russell was given knowledge about healing himself, and while still dead, Russell did heal himself. He awoke fully healed, amazing doctors and family, not to mention the undertaker. During the near-death experience Russell became aware of his life's purpose.

In 1921, at the age of 50, another experience changed Russell's life again. He entered the "Universe of All-Knowing," where God kept him for thirty-nine days and nights. He received divine knowledge about the essence of All-Knowing, and he wrote down every word, without one word change or even an erasure. He said they were God's words, written through him, to form a foundation for the new civilization. A number of books resulted from this experience, including *The Universal One* and *The Secret of Light*. Jesus had said, "God is Light" but no one knew what he meant. Russell said, "In these thirty-nine days and nights God spoke to me continuously in the language of Light which I knew as well as I know the English language . . . ." Russell said, "It necessitated the uplift of the whole human race with new comprehension of the New Age. The seed of that new comprehension for the uplift of the human race was mine to sow . . . ."

Russell waited nearly five years before writing his first book and giving it to the world. Although scientists either ignored it or laughed at it, gradually some began to comprehend what he had written, and two Nobel prizes were awarded to scientists who adopted ideas that had originated in that book.

Russell received an honorary Doctor's Degree of Science and was elected president of the prestigious Society of Arts and

Sciences. He presented a lecture on hydrogen to 400 of the most distinguished scientists in the world. Scientists had thought hydrogen was a single element form, but Russell told them there were six other full-tone elements of hydrogen, a whole octave of it. From that lecture, scientists discovered heavy water. He also foresaw the discovery of the elements neptunium and plutonium.

Russell said he was sent by God to be a messenger. In his 1921 experience, Russell was told about the upcoming world situation. In a 1946 lecture Russell said, "The world had not been prepared to accept or even comprehend the new cosmology until now, nor has it been willing or ready to accept the New Age philosophy. The world needed to suffer in order to understand the simplest of principles, the unity with man to God. The world of men had to reap the harvest of the hate, selfishness, and greed it had been sowing for centuries . . ." This date was written down by him in 1921.

Russell explained how all this knowledge had come to him. "I will put it very simply. In May of 1921, God took me up into a high 'mountain' of inspiration and intense ecstacy. A brilliant flash like lightning severed my bodily sensation from my whole consciousness, and I found myself freed from my body and wholly in the Mind Universe of Light, which is God." And then God said to me, "Behold thou the unity of all things in Light of Me, and the seeming separation of all things in the two lights of my divided thinking . . . ."

"And the secrets of the universe were unfolded to me in their great simplicity as the doors to the Light opened fully to my consciousness. In less time than it takes to put into words, I knew all there was to know, the cause and effect." He went on to say, ". . . I at once had the key to all science, mathematics, chemistry, astronomy, and mechanics, and likewise to all underlying principles of creation; of life and the healing principle; of continuity in a universe in which there is no death . . . .

"And the Law was mine to know, the One Law which governs all things, extending from the Source through the universal pulse beats which motivates all things and it was made known to me that I must extend knowledge of this Law unto all humans relating to help remake the world in its new day which God has planned. And so it happened that I, who never had any school or university training above the primary grades, thus knew instantly, while in Light, what all the universities in the world could never teach. . . . All knowledge exits in the Mind Universe of Light, which is God. . . . That all Mind is One Mind, that men do not have separate minds, and that all knowledge can be obtained from the Universal Source of All-Knowledge by becoming One with the Light Source."

Walter Russell went on to establish the University of Science and Philosophy in Waynesboro, Virginia, and died in 1963.

The Mind

Russell's Divine Source told him that Mind is all that is or ever will be. It is the seed of the universe, and within this seed is the whole of the universe. **Mind is a material substance which has no beginning, no ending, and no boundaries.** It is intelligent, powerful, and present in everything. In quantum physics this is the quantum field, the Mind Matrix.

The quality of mind belongs to motion caused by thought. What causes the appearance of form of the One Mind Substance (Universal Mind) is a change of state caused by motion through thought. The Universal Mind itself does not change. Russell wrote that it is the illusion of motion that creates the illusion of dimension in the material world. The Universal Mind is the one pre-chemical substance from which all elements and compounds of the elements make their appearance. They have no existence except the appearance of existence, and therefore are only illusions.

Form is not an attribute of the Universal Mind and has no existence other than appearance. Form is an attribute of motion only, just as bubbles whirling in water are an attribute of whirling water and not the substance of water itself. When motion ceases, the substance of water remains, but the bubbles disappear.

Physical appearances register the action of the thinking process. These physical manifestations all eventually disappear back into their Source, which is Universal Mind Substance, an absolute timeless state of motion-in-inertia. Inertia is defined by Webster as "a property of matter by which matter remains at rest or in uniform motion in the same straight line unless acted on by some external force." That external force is our thoughts, according to Russell. Inertia is that state of motion which continues without force. All energy is caused by a disturbance of inertia. In other words, the appearances of matter disappear back into the ocean of Universal Mind, motion-in-inertia, and become part of the void until acted upon again by the force of thought.

In quantum physics, light waves become light particles when an observer observes. Simply looking at something as tiny as an electron changes its properties. The act of observation is an act of creation with consciousness doing the creating. This material Substance of Mind (quantum field) is an all-pervading ether which is indivisible, inseparable, indestructible, and unchangeable. Indian philosophy calls it *prana* and the Chinese call it *chi* (qi). In the West the quantum field is called life force or subtle energy. Scientists call it zero point energy. Quantum physicist David Bohm calls it the implicate order. **Universal Mind Substance contains the appearance of all these dimensions of separation, form and events, in their states of motion that have registered the dynamic process of thinking.** The words "spirit" and "ether" are used to describe the dimensionless universe, while solidity is used to express the compactness of the apparently measurable universe created by thought. The Universal Mind Substance is a spiritual substance. Spirit is Light.

Thinking is the only energy of the universe. (Quantum physics demonstrates this with the involvement of the observer.) This force called thinking compels mind into concentration and de-concentration in sequence. **There is no other energy but thinking! The universe is Mind only.** The universal constant of energy caused by thought registers in the Universal Mind. Illusions, or the material universe, are caused by the thinking mind. The process of thinking carries throughout the entirety of the universe at an incomprehensible speed in waves of light units that return to that exact center where they originated. (Nonlocality). **Thinking is an action followed by a**

reaction to that action. Every microscopic point in Divine Mind becomes the center of the universe of Mind.

States of motion in opposition, caused by thought, bring about the appearance of change from the state of inertia in Universal Mind into the appearance of separation into parts. Such effects create dimensions and can be quantified such as heat, cold, mass, gravity, etc. **Anything that can be measured is only an illusion because it all goes back into the ether of Universal Mind.** (This concept mirrors David Bohm's explanation of the explicate order going into the implicate order.) These effects are not a change in Universal Mind, but only in the dimensional world. All effects have the appearances of dimension. **The cause of all effects is dimensionless. Cause is existence.** Universal Mind Substance, light waves, matter (light particles), and energy are externally existent as cause and should not be confused with the effects in motion which are dimensions and not existent. These effects are illusions of existence. In other words, changes of state appear to change the character of the Universal Mind, but appearances have no existence.

Humanity is accustomed to appearances because we live in a universe of relative appearances. It is impossible for appearances to exist without relation to other appearances. Without the illusion of separability, space could not exist. Without events, time would not exist. Without motion in opposition, there would not be heat, cold, color, sound, sex, mass or any of the other effects of thinking. Man's material world is but a variance in state of motion of the Universal Mind Substance (Mind Matrix) caused by thinking. Material objects appear to have separate and distinctive characteristics of their own such as color, weight, shape, etc. in varying degrees. They appear to be separate and different, but they are not, because they are manifestations of the same One Universal Mind. The appearance of separability is due solely to periodicity of motion in opposition (action and reaction) caused by thinking. Man has the potential to change any one state-of-motion into any other state-of-motion through his thinking mind. In other words, man has alchemic abilities to change one element into another.

The Universal Substance of Light is a material substance of variable motion. It is caused by two apparently opposing forces of action and reaction, which constitute the thinking process. Light is without form, but potentially contains all that man calls form. It is nothing but an appearance, which is an effect in opposition. The more the opposing forces are caused by thinking, the greater is the rigidity of form and mass. (In quantum physics the observer's mind causes light waves to become light particles.)

Evolution is the concentrated desire of the mind to create the idea of many substances through mind out of the Universal Mind. Dissolution is the desire of mind to return that which is created back into its Source.

Thought

The Universal Mind is a formless thinking substance, the quantum field or Mind Matrix. Without a thinking Universal Mind Substance, creation would not occur. God is an ecstatic thinking process. **Thinking is a process that causes all motion in the Universal Mind.** It is purely a mechanical process that is periodic in its evolution, through one cycle after another without end. Thinking registers the effect of its passage through the ocean of the Universal Mind. A myriad of rotating particles in the Universal Mind Substance registers the thinking mind. All the effects of motion are orderly and periodic.

Without the force of thinking throughout the tranquil Universal Mind, there would be no variability in the Universe of Mind. There would be no form. Stars, planets, animals and plants are all appearances of whirling forms born of the thinking mind.

The process of thinking is a swing of the cosmic pendulum out from the stability of Universal Mind Substance to apparent instability and back again to stability. The registration of this effect in the ocean of the Universal Mind is contained within the form of an elongated sphere. All its variabilities will be imprinted upon two exact spiral waves within the two halves of the sphere. Light, electricity, magnetism, sound, sex, life, death – the soul of all things – and all that man can comprehend originate in these two opposing waves caused by thought, action and reaction.

These two opposing spiral waves within the sphere represent the entire process of thinking. Everywhere throughout the One Universal Mind Substance is an agitation that is caused by the orderly process of thinking.

The action of thinking constitutes a series of events in sequence. Intervals between events in sequence constitute the effect of motion called time. Without a sequence of events, there would be no time. **Time originates with the act of thinking.** It is an effect that appears with motion in opposition caused by thinking and disappears into motion in inertia (Universal Mind).

Events in sequence give birth to the appearance of time, and the opposing pulsation of light. The opposing pulsation gives dimension and form that man calls life. For man, there is no life without form. However, form and dimension are fleeting. **Only life is eternal. Life is merely the action of thinking.** Thinking is as eternal as the thinking Universal Mind is eternal.

Universal thinking is rhythmic thinking. The entire Universal Mind is thinking in variable but orderly rhythm. All thinking is expressed in measurable and opposing impulses of motion. All motion is action and reaction, which are the opposing impulses of motion. **Thinking is the cause of action and reaction. The action and reaction are illusions.**

The Process of Thinking

According to Russell's Divine Source, **the electro-magnetic opposition mind as expressed in the process of thinking is the source of all the energy of the universe.** The action impulse and reaction impulse of the thinking process constitute the appearance of motion in opposition. The action impulse is the generation, and the reaction impulse is the radiation of the process. Generation is electric or positive, and radiation is magnetic or negative. All motion is in equilibrium, whether in opposition or in inertia. Thinking is the electric generative action and the magnetic radiative reaction of the mind. Thinking is a simple process with very complex effects. Motion is continuous and thinking is continuous.

The universe breathes as does man. The process of exhalation and inhalation in sequence is a characteristic of all phenomena of matter. Every light particle, atom, and molecule breathes.

Thinking is registered in matter. In other words, matter registers the divine idea of thinking mind. That is, light waves become light particles. As a person concentrates electrically in his thinking, the idea becomes more and more distinct in form. The form remains distinct as long as man concentrates dynamically upon thinking that idea. When concentration diminishes, the idea becomes nebulous in form.

The opposition of the two opposing forces (action/reaction) causes opposite impulses that form a spiral wave along which thought is registered. This registered idea of thinking is expressed gravitationally in what man calls the elements of matter. All elements of matter are generative and they are also radioactive. Generativity builds the elements and radioactivity breaks them apart. The elements are composed of separate particles in motion called "light units" or "light corpuscles." These genero-radiative waves are the medium for reproduction of ideas throughout the entire universe. All ideas of mind produced by thinking are reproduced throughout the entire universe of Mind and are measured in waves of electromagnetic opposition.

All motion is spiral and all direction is curved. All the elements of matter, in all states of motion, are but orderly and periodic variations of the one substance, Light. All the elements of matter are spread out progressively on ten octave waves, and the seven tones of each octave. All solids of matter are but integrated light. (Note: In his book *The Universal One,* Russell speaks of ten octaves and seven tones. Later books tell of nine octaves and eight tones.)

Appearances

Matter is an appearance and has no existence except as an illusion. During man's evolution in a universe of motion caused by thinking, the complexity of appearance has caused man to form inaccurate concepts of reality and unreality. Appearances have become facts and the reality behind appearances only conjecture. Man's unreality is spirit or etheric things.

Man needs to alter his reality concept to the reality of Mind as the source of the illusion of solidity, writes Russell. **The only reality is the Universal Mind or Mind Matrix**. Form is an appearance of reality in a lower octave of the material Substance of Universal Mind, of which man has control within the limitations of mind. Matter, created by light particles, is the substance of thinking mind and man needs to learn that he is Mind. **Mind is omnipresent, omniscient, and omnipotent.** Until man learns he is Mind, he will be a slave of the illusions of Mind. Man instead could be the master and creator of the illusions.

The Polarity Principle

The sex principle or polarity is the controlling cause of both form and motion. Without polarity, neither form nor motion could continue. **Sex is also the motive behind force and motion**. Polarity is the division of the Father-Mother, the Universal Mind, into apparent opposites. The division is due to the opposite desires of electricity and magnetism, which are expressed in the action and reaction of the thinking process. Sex is the active desire of mind for division by polar opposites and its reactive desire for unity. The motive force of this polarity demands separability into two, and equally desires union of the apparent two into one.

Nothing can manifest without desire. Desire dominates all thinking, and thinking desires all matter. All desire is sex desire, which is the force behind it. Basically, the sex form is the ripple of the apparent many things that arise out of the reality of the ocean of the One Universal Mind Substance. **Matter is unable to continue its appearances of existence as form without the desire to continue such an appearance.**

Polarity (sex) is an electromagnetic equalizer of matter in motion. It is the division of the One Force into electricity and magnetism. These two opposite polar forces are really two pulsations of the One Force. Polarity is the division of all things into opposites, action-reaction, male-female, electro-magnetic, generation-radiation. However, sex is evolutionary in all things as everything is both male and female in the Universal Mind.

Polarity first asserts itself in form by thought. All growth is impelled by force and motion that is controlled by polarity.

There can be no force without thinking, and there can be no idea of mind without polarity opposition. Without sex union in mating, the idea of any one thing would be extinct as an appearance. In every seed of an idea lies the whole of the idea. The seed of all ideas is thinking. This is the Law of Evolving Existence.

Units of light, electricity, magnetism, sex, heat, etc., are all male and female. Each unit is predominantly male or female, positive or negative, which generates or radiates. Each unit is variable and becomes predominantly male (electro) or female (magnetic). Their variability is orderly and is governed by the law of periodicity or the law of motion. The opposing electromagnetic impulses are polar expressions in creative thinking. All thinking is rhythmic thinking and all thinking is polarity thinking.

Creation is transformation of the Universal Mind by rhythmic thinking of idea into apparent opposites and into that which man calls form. Creation is nothing but a concept of mind. It is an illusion which is an effect of thinking. All thinking is registered in Light, and all light is polarity expression.

Russell writes that **Light is the language and energy of thinking** (as demonstrated by quantum physics). All energy is the energy of sex expression (desire) caused by thinking. There is no other energy. Through evolving polarity energy, light causes the transition of the Universal Mind Substance into form. All form is produced by the male, electro-positive actions of the charging electro-oscillation of the universal life principle caused by thinking. It is undone by the female or negative reaction of discharging magnetic-oscillation caused by thinking.

Reproduction

Light, created by thought, is the power God uses to create the appearance of form. For man, light is power to use in perfecting the body. Man's body is an individual effect of the whole of man and is the product of thinking, writes Russell. **The universe is the product of God's thinking, and God's thinking is man's thinking.** Light is God's medium for expression.

Form originates from the desire to express idea. All form is a form of polarity. Form is born anew with each impulse of thinking. Man thinks an idea. Thinking is the force of motion and all motion is expressed in waves. All waves of motion are both male and female. Action is male and reaction is female. Thinking produces form from the Universal Mind Substance which is electro-magnetic, male-female, action-reaction. Evolution is male or growth which is electrically preponderant and devolution is female or magnetically dominant. Man can only mature an idea through orderly periodicity of thinking.

The reproductive force is the union of two opposites, male or female and positive or negative, into the constant of one which is in equilibrium. Reproduction means the repetition of a state of motion. **All phenomena of nature are repetitions, and all ideas of mind are repetition.** Our universe is a repetition of motion. It is a universe of cause and effect. **All effects are caused by thinking and registered in motion**. That which is produced must be reproduced. All states of motion are forever reproduced. The reproduction of all ideas is the result of the action (male) and reaction (female) which registers that idea.

Every action has its equal and opposite reaction. The action and reaction are not two, but one when their energies are combined. Perfection in mating lies in the union of exactly equal and opposite male action and female reaction. There is no reproduction without union.

Stable union will always reproduce true to species, but unstable union tends to return to a separate tonal state. If a mate is in an unstable union and finds a more equal mate, it will always go to the more equal mate. That which is true of chemical union is also true in all species of life.

All motion is expressed in waves but those waves are not undulating. One whole wave is not complete before another wave begins Waves are effects which evolve as their cause evolves. Waves reproduce themselves part for part in sequence throughout the universe. Light as a wave reproduces itself; it does not travel. Light from the stars or sun that man sees has never left the stars or sun. Man sees only the reproduction of the light.

All energy expresses itself in motion. All motion is expressed in waves. All dimensions are measured within waves, and therefore all dimensions must be measured within waves. To know the wave is to know the secret of creation. The wave is nature's method of transferring the dimensionless concept of form in inertia of the Universal Mind to dimensionless form in man. This happens by an impact of the thinking energy of the concept against the Mind Substance. The purpose of the wave is to transfer states of motion from one potential to another. All waves are the expressions of a force, a record of the universal constant caused by thinking or its multiple in accumulated energy. Waves are simultaneously generated from inertia of the Universal Mind at intervals.

This principle also applies to ideas of mind which originate from thinking. Ideas have no place or position in time or space, as all ideas are universal. The state of motion which represents the idea is reproduced as light. The law of reproducing idea is the principle of universality. **All that is, is of everything else that is.** The spiral generative-radiactive waves are the medium of reproduction of ideas throughout the entire universe. Matter or light particles register the divine idea of thinking mind. There can be no idea of mind without polar opposition. No idea of mind, once begun, will ever end. Light particles organize and integrate into complex forms of matter. These elements of matter, light particles, register the idea of all thinking. All ideas are energy. The energy of one idea can be transformed into the energy of another to continue the idea of that other. The principle of reproduction is the principle of energy transmission. All ideas continue by the energy of thinking. The great Law of Evolving Idea is evolutionary and continuous, as is the recording of idea. The form of an idea is stored in inertia of the Universal Mind. Both idea and form of idea return to motion-in-inertia (Universal Mind Substance) and remain there for a time of formless idea. All concepts are but recollection of eternal ideas.

Energy transmission is another polarity, conduction and induction. Induction is the raising of a low potential away from inertia. This is electro-positive or generative. Conduction is lowering of high potential toward inertia which is electro-negative or reaction. That which is generated must be radiated.

All mass is regenerated by absorption of the impacting radioactive energy of all other masses. All mass is degenerated by its own radiation. All mass is generated by accumulation of the universal constant of energy into higher potential. One opposite is always born of the other, which in turn becomes the cause of the first. As a result, **no idea of mind ever ends.**

The Finite Universe

To reiterate, mind is all there is. It is limited in its power, knowledge, and imagination. Imagination is its creative force. Every effect there is, is caused by the mind. These effects are measurable and have the appearance of dimensions. All dimension is limited and therefore all effects must be limited. It is impossible for mind to create infinite complexity. Mind cannot go beyond the ten-octave cycle of cause and its included effect (see next chapter). All effects evolve to their limit and dissolve back again into cause without any change taking place in cause.

Man believes that a finite universe is inconceivable to the human mind. Einstein said the universe is finite but wrongly said it has form. He rightly said that any light of direction in space is curved. It would ultimately come back to its origin. All effects of motion are actions which must end where they began. Therefore, direction must always be curved. These lines curve in a spiral.

There are no parts or places to the Universal Mind or Mind Matrix. Every place is the same place. Any disturbance in the Substance of Universal Mind returns itself to its exact beginning. The disturbance radiates in a sequence of reproductive waves, which give appearance to another and nonexistent effect that we recognize as time and form.

Thinking is universal and all its effects are simultaneously universal. (Quantum physics calls it entanglement.) All motion is simultaneously universal in this universe of equilibrium. Sequence is an appearance because of form. Without form, we would not have the experience of existence. **All form has no existence.** It is a question of gravitational intensity.

Time is an effect of gravity and also has no existence. Form, time, and space do not exist. Neither do position or dimension. Unrealities disappear into the realities of Light. The Inner Mind of man, or Higher Self, thinks in Light, independent of the illusion of the thinking outer mind which is ego. By understanding the reality of mind, the potential for changing our personal reality with our thinking mind is unlimited.

Summary

1. Mind is all that is, a material substance without beginning or end.
2. Mind is intelligent, powerful, and ever-present.
3. Thought causes motion in the Universal Mind Substance.
4. All elements and compounds originate from Universal Mind.
5. The appearance of matter, which is an illusion, disappears back to the Universal Mind Substance.
6. Anything that can be measured is an illusion, including matter, space, and time.
7. Universal Mind contains the appearance of separation into parts.
8. The appearance of separability is caused by thinking.
9. Light is the Universal Mind Substance caused by apparently opposing forces of action and reaction, which constitutes the thinking process.
10. Form is an attribute of Universal Mind and has no existence other than appearance.
11. The One Universal Mind is a formless, thinking substance.
12. Thinking is a process that causes all motion in the Universal Mind.
13. Everything is caused by thought.
14. Thought produces two opposing spiral waves, electric and magnetic.
15. Thinking is the source of all energy in the universe.
16. Thinking is an action (electro-generative) followed by a reaction (magnetic-radiative) to that action.
17. Life is merely the action of thinking.

18. Matter registers the idea of thinking mind.
19. All ideas of mind are reproduced throughout the entire universe and measured in electromagnetic opposition.
20. All elements of matter are spread out on ten octave waves and seven tones of each octave.
21. The only reality is the One Universal Mind Substance over which man has control.
22. Mind is omnipresent, omniscient, and omnipotent.
23. Polarity (sex) is the controlling cause of motive, form, and motion.
24. Nothing manifests without desire.
25. Polarity is the division of One Force into electricity and magnetism.
26. Each unit of light is predominantly male or female which generates or radiates.
27. Light is the language and energy of thinking.
28. Light causes the transition of Universal Mind into form.
29. Form is produced by the male, electro-positive action of thinking.
30. Form is destroyed by the female, magnetic reaction of thinking.
31. God's thinking is man's thinking.
32. All waves of motion caused by thinking are both male (action) and female (reaction).
33. All ideas of mind are repetition, and all natural phenomena are repetition.
34. All effects are caused by thinking and registered in motion.
35. That which is reproduced must forever be reproduced. No idea of mind ever ends.
36. Light reproduces itself; it does not travel.
37. All that is, is of everything that is.
38. Thinking is universal, and all its effects are simultaneously universal.

Chapter Eleven

THE THINKING MIND

A New Paradigm on Thinking

"Man must think in Light to truly comprehend the Mind," wrote Walter Russell in *The Universal One*. The ability to think in Light is a power found in every atom of man's composition. Man currently thinks only in the lower wave octaves and is unaware of the higher octaves within him. Man has the ability to shift his thinking from the low speed of the outer mind to the higher speed of the Inner Mind at will. Man's outer mind (ego) is an integration of light into the lower octaves of creation. The outer mind thinks in dimensions or illusions of mind. Remember, anything that has dimension is an illusion. Mind, matter (light particles), light waves, and energy are all the same. They are one, constant, and eternal. However, the dimensions of mind, light, and energy are illusions of Universal Mind. (Oneness, Universal Mind, God, Light, Mind Matrix, quantum field, implicate order, Mind Substance, and unity consciousness are words to designate the same concept. Form, illusion, dimension, time, space, matter, and explicate order are also the same.)

The Universal Inner Mind (Higher Self) thinks in Light and registers that thinking in Light, which is integrated into the idea of thinking. Universal Mind, Matter, Light and Energy are externally existent as cause. They give rise to the outer mind of man and should not be confused with its effects in motion which are dimensional and therefore nonexistent. Dimension, such as form and events that can be measured, is nothing but objective outer mind thinking. Although, the Universal Mind Substance is eternal, illusion of dimension and form is temporary and not real.

Man's Inner Mind is Light in the highest octaves consisting of inspiration and ecstatic thinking. To man's Inner Mind, Light is always Light and never disappears. The Inner Mind (Higher Self) knows only Light and not darkness. With Inner Mind, matter is dimensionless. The Universe of Mind Substance is a dimensionless universe. However, the universe of the idea of Mind caused by thinking is a universe with dimension. The creating universe is the idea of Mind. It is the divine concept of motion caused by thinking and the apparent separation of One Thing into the appearance of many individual things.

Dimensions

The divine concept of a universe of form is the idea of the image-making faculty of Mind. It is an illusion of divine imagination. Mind thinks the idea of mind into an illusion of form. When form appears, so does dimension. Form is the result of motion generated by the concentration of thinking. Creation is but an illusion as a result of a complex state of motion.

All motion is comprehensible and caused by thinking, but controlled by the mind. All motion is measurable, and measurements are dimensions. Dimensions are intervals that define the relation of one illusion to another. Illusions of form are varying states of motion locked into the separate states of frequency which define those illusions. All states of motion, no matter how complex, are reduced to frequencies of ten-wave octaves and seven tones plus mid-tones and their master tone.

All of creation can be reduced to 140 elements which comprise the entire complex of God's thinking. The elements of matter are composed of all the same "substance" in varying degrees of motion. These elements differ from one another by the relationship of their interval in motion or measurement of their separate dimension. The divine concept is not a universe of motion but an idea in form. Motion, an attribute of thinking, is necessary to produce the form of an idea.

There is no substance which nature can produce that man

cannot also produce when he knows the dimension that measures the illusion of form. Man has control of more energy than is necessary to transmute one element or compound into another. All that he lacks is the knowledge of dimension.

According to Russell's Divine Source, there are eighteen dimensions characterized by an orderly periodicity. Beyond these eighteen dimensions, there are no measureable intervals of motion.

The eighteen are:

1. Length
2. Width
3. Thickness
4. Duration of Time
5. Sex
6. Pressure
7. Potential
8. Temperature
9. Ionization
10. Crystalization
11. Valence
12. Axial Rotation

13 Orbital Revolution

14. Mass
15. Cold
16. Plane
17. Tone
18. Ecliptic

The spectrum of the elements has been sorted into 7,000 colors of light. The 7,000 lines of light make up the letters of the universal language of Light. This is the language by which God speaks to man. Once man knows the language, he will think in it. **When man thinks in Light, he will know he is the son of God and one with God. Man will know he is Universal Mind. He will also know that he is omnipotent, omniscient, and omnipresent.** Man will know he is eternal life, and that he is one with everything. Within God's alphabet of Light, the master key of knowledge opens the door to man. Mind creates and controls the energies of the elements. Our mind can rearrange these energies and combine them in accordance with its desires.

Russell said there are ten octaves with seven tones in each octave for each dimension, except for the dimensions of mass and tone. Each octave is divided into seven tones of energy, and the ten octaves constitute one cycle. All eighteen dimensions can be measured by the same octave periodicity. Periodicity means the orderly progression to and from the maximum effect of each octave. Mass accumulates all down the entire octaves, and tone lowers from the highest note of the cosmic keyboard to the lowest. The thinking mind has devised orderly progressions as its method of evolving the ideas of universal thinking into the appearance of form.

A cosmic clock can be used as a metaphor of the ten octaves. The seconds in the cosmic clock can be viewed as light particles created by the thinking mind which make up the atomic structure of form. The mainspring of the cosmic clock is the ten-octave cycle. Five of these octaves are decelerating the time dimension and accumulating power. The other five octave dimensions are accelerating time and releasing power. Five octaves represent inhalation and winding the clock and the other five represent exhalation, the unwinding of the clock. The cosmic pendulum ticks off the varying dimensions of all motion forever and ever.

All energy equalizes in two equal, opposite swings of the cosmic pendulum. The cosmic pendulum swings between positive (electric) and negative (magnetic) caused by the thinking mind. In doing so, it transfers its constant of energy from one dimension to another. Russell describes positive electricity as producing gravity by compression, which results in generation. **This is a centripetal force produced by gravity that compresses thought waves into body forms.** Negative electricity or magnetism produces the condition of radiation which degenerates. **It is a centrifugal force that causes radiation and expands thought-waves to void form.** All energy is caused by a thought disturbance in inertia of the Universal Mind. The energy impact of thinking against inertia is the cause of motion that initiates the cosmic clock and cosmic pendulum. The energy constant of thinking that initiates the motion of Universal Mind is the cause that is registered in wave form. It is registered as an idea in the Mind Substance that develops into form, depending on the desire.

Light and Ideas

The Universe of Mind (Mind Matrix) is a dimensionless

reality. **This Substance of Mind is Light (Light waves) and Matter (Light particles). The electromagnetic energy of thinking is the force.** The inspired purpose of the creating physical universe is to conceive idea and give it the appearance of form. This is done through imagination, the image-making facility of mind. Form and dimension are attributes of motion, not attributes of matter. The effects of motion constitute the creating universe. The effects are an illusion, an idea, which has its cause in the thinking mind.

The universe of illusion is a corpuscular universe. All dimensions of matter and energy are corpuscular. These corpuscles are small particles of matter or light, the Universal Mind Substance. This Mind Substance has been churned into motion by the electromagnetic forces of the thinking process. The motion complexity of the corpuscles is the only cause of illusion. The Universe of Mind is what man has wrongly characterized as the spiritual universe and the universe of dimension as the physical universe. **The Universe of Mind is cause and real. The physical universe is an effect and an illusion.**

When the range of man's sense includes the entire range of Light, man will then know that there is no difference between the physical and spiritual, except for a variance in motion. Variance does not constitute a difference in substance. From the high octave of spirit (light waves) to the low octave of matter (light particles) and back again to spirit is a transition to nonappearance from appearance and back. (From the implicate order to the explicate order and back.)

Light is corpuscular. It does not travel, and the appearance of its traveling is another illusion. All light units of matter are given the appearance of form by the magnetic reaction of the electricity's attempt to seek high pressure. Essentially, light freezes and becomes crystalline. All matter is crystalline, the apparent solids of light. Man's solids are but light retarded in its 12th dimension of axial rotation. It is sustained deceleration of motion by the energy of gravitation sturdily opposing disintegration.

The substance of matter is eternal but the appearance of matter in form is temporary. That which is not eternal will eventually disappear (explicate order to implicate oder) until the swing of the cosmic pendulum causes an orderly and periodic reappearance (implicate order to explicate order). Nothing can change but can only appear to change.

Transition is a journey, from visibility to invisibility. The journey too is an illusion. Creation is a journey of the thinking mind of the Universal One into the illusion of the divine idea. Creation is but the materialization of image from the thinking process of mind in action. The sole purpose of mind is thinking, and ideas are the sole product of thinking. If the Substance of Universal Mind had not been a thinking substance, the universe would be without the drive of opposites, without sex or motion, which makes appearances and effects possible. If the Universal Mind suddenly stopped thinking, all created things would instantly disappear. The created universe of illusion would disappear.

Each newly-born corpuscule of light is a center of disturbance in the Universal Mind to which the entire universe will respond from its alternating electromagnetic pulsation. Once the motion starts, it will not stop until it runs the gamut of the ten-octave range, and then it will begin again. All the effects of thinking are omnipresent in the universe. There is nothing which is not universal (Entanglement is discussed in Chapter One).

Light corpuscles constitute created things caused by thinking. **Creating these light corpuscles is as continuous as thinking is continuous because thinking creates light corpuscles.** The reproduction waves, electromagnetic motion in opposition, are made of the corpuscles that register in the Universal Mind as ideas of the thinking man. These corpuscles are living and thinking beings that are male and female. They organize and integrate into complex forms of matter, which register the idea of all thinking. In other words, light units are born into the universe of illusion as an appearance, and register the idea of thinking mind. Man, himself, is a complex organization of light corpuscles that function under the idea of thinking.

All light corpuscles are both generative and radiative, but each is predominantly one or the other. They integrate into form and disintegrate from form. They have periods of sleep and wakefulness. Repetitiveness is an absolute trait of all effects of motion. All mass inhales and exhales. Inhalation continues the apparent existence of form and matter, and exhalation is dissolution of matter. The substance of thinking mind in action is light. Thinking registers itself in light which integrates ideas into form. All idea is energy, and the energy of one idea can be transformed into the energy of another to continue the idea. One light unit does not destroy another by absorbing it, but assimilates the idea of the other. Idea continues itself as idea by the energy of its thinking.

Light units grow, just as all things grow, by the absorption of other light units. In this way, man evolves. Each thinking light unit adds to the idea of itself from the source of idea and all thinking registers itself in new light particles.

These light units, assembled as man, think individually but coordinate as a unit. Each will function in its place or the idea of man could not continue. The individuality of man as a part of the universe is analogous to the individuality of a light unit as a part of man. This is but the great Law of Evolving Idea. Idea is evolution and continuous as is the recording of idea.

Electricity and Magnetism – Polarity

All mass is both electric and magnetic. All electromagnetic mass forms into systems of units which revolve in spiral orbits both centripetally toward the nuclear center and centrifugally away from the nuclear center. Electricity is inhalation while magnetism is exhalation. Electricity is the positive half (male) of the One, while magnetism is the negative half (female) of the One. They become One with nonopposition.

Electricity is an attractive force within mind. It compresses the noncompressed Substance of Universal Mind into forms created by thinking. It evolves those ideas by raising the potential of energy by accumulation. Magnetism is the repellent force within the Substance of Universal Mind. It expands the forms created by mind in the thinking process. It causes dissolution through lowering the potential of energy by releasing the accumulation.

Electricity is the force of gravitation that moves spirally inward toward the nuclear center of its mass, while magnetism is the force of radiation that moves spirally outward toward its extreme. Each of these forces balances itself as well as its opposite force. The generative force of electricity decelerates motion in order to accumulate mass, while the radiative force of magnetism accelerates motion to dissipate. They are two different dimensions of motion.

Electricity is the basis of the atom. Within every atom of the Universal Mind is all the power of the universe. It is dynamic. Magnetism is born of electricity. The apparent opposites of electricity and magnetism merge to make one. Electricity is magnetism revitalized while magnetism is electricity devitalized. Action and reaction when added together are but one. The apparent opposite of electricity and magnetism is the cause of creation. The active electric force, the Father source, is the creative force of the universe while the reactive magnetic force, the Mother force, is the degenerative force. The Father-Mother force unite in the reproductive force.

Electricity gives form to an idea while magnetism gradually reduces that form back to idea. Form is temporary, as it appears and disappears, and is periodically repetitive. **An idea is the eternal record of form or the soul of form.** Death is the disappearance of the form itself, but not the idea of form. All evolution of an idea of mind into form is electric. Youth corresponds to the first five octaves of accumulation, while aging corresponds to the last five octaves of release.

Electricity and magnetism are opposite principles of the quality of mind, allowing it to be a thinking substance. Thinking is nothing but electromagnetism registered in Light. It is both energy and matter, all one and the same. Electricity desires a universe of form and dimension. Magnetism desires a formless and dimensionless universe and is that force of Universal Mind that tends to preserve the Oneness of universal uniformity. The electricity is the force that creates the illusion of form, while magnetism prevents the apparent separation of Divine Mind into its parts. Its force is used to destroy the illusion. Electricity and magnetism are the major dimensions of the universal constant of Mind. The attempt to divide the One Substance of Universal Mind gives the appearance of these two substances. It develops two equal and opposite states of motion or forces that create the illusion. Electricity and magnetism are illusions of the Universal Mind.

Electricity has no existence and belongs to motion, not to substance. Electricity acts upon the desire of mind to create. Ideas and form must be developed out of the Universal Mind Substance. Both idea and form of idea return to motion-in-inertia Universal Mind (implicate order) and remain there for a period as formless idea. The union of electricity (Father Force) and magnetism (Mother Force) then recreates the form of idea.

Formation of Matter

The elements of matter come about through aggregations of corpuscular light units gathered together in systems of atoms. **Atoms differ from solar and star systems only in dimension.** A bigger atom such as a solar system is not different from an elemental atom. Matter progresses through cycles of appearances and disappearances. (Visionary physicist Nassim Harramein has mathematically shown that the subatomic world obeys the same laws of gravity and electromagnetism as the macro-world of the universe. He gives mathematical evidence to Russell's revelation that the subatomic world reflects the solar system in miniature. It is the author's prediction that once Harramein's findings have been replicated, he will receive a Nobel Prize in physics for discovering the mathematics explaining the elusive Unified Field Theory of Einstein, the Holy Grail of physics. Harramein is associated with The Resonance Project Foundation in Houaloa, HI.)

Every pulsation of thinking mind brings particles into appearance and then disappearance. As all other forms of life, matter is evolving and devolving. An iron particle is not always an iron particle. It changes its dimension in an orderly manner until it has run the gamut of all elements of every octave and is returned to its own octave. A sodium particle was once a light unit of hydrogen. It becomes everything else in the universe. It is not the substance which determines its effect as an element but only its position and its dimension in its octave. There is only one Universal Mind that appears to be divided into many substances, known as the elements of matter.

The One Universal Mind Substance first divides itself into the appearance of two opposite states of motion, which registers as positive and negative elements. Then from them two more opposite elements appear. From each division two more appear of increasing potential and crystallization. One element is formed which registers itself in the physical realm. Every element in the entire periodic table is a transmutation from the proceeding element of its cycle.

All states of motion are measurable and all are under control of mind. By drawing from knowledge of causes, man can change any state of motion into another state of motion. As a result, he can transmute any one substance into another in an alchemical process.

The created universe of matter is a consequence of mind concentration upon the idea of form until the idea of form disappears through mind deconcentration. The exact energy of thinking is registered in the electro-positive charging system, and the reaction of the action is registered in the electro-negative discharging system. The ten-octave cycle integrates and disintegrates light units into the atomic system we call "elements of matter." The universe is the sum total of electric actions and reactions expressed in thought waves of two-way motion.

The Ten-Octave Cycle

The ten-octave cycle begins and ends with an absolute equalization of all dimensions of motion. Creation is continuous. Its beginning is its ending. As a matter of fact, there is no beginning or ending. The effect of electricity to accumulate power at the expense of time is continuous throughout each tone in the ten-octave system. Each octave is a separate and complete sequence of inhalation and exhalation.

During the first five octaves, the power dimension gradually conquers one time dimension. These five octaves are male and electro-positive. They reflect the direction of birth, youth, and growth. The second five octaves are electro-negative, similar to death and dissolution. The cycle of the elements evolve from low to higher and devolve to low potential. The entire ten-octave cycle is an accumulation of the constant of energy into higher-power potentials at the expense of speed. Speed means the quickness with which a state of motion can reproduce itself. **Thinking is the only source of energy.** Enormous power is accumulated by thinking the idea of the universe and registering that idea in repetitive systems.

The breathing of the Universal One is electromagnetic. All mass is life and all mass breathes, assimilates, and eliminates the life-giving property of that breathing. The inhalation action is the cause of electricity and the principle of growth and evolution, the male principle. Exhalation is the cause of magnetism, dissolution, the female principle. This electromagnetic oscillation of the universe is the heartbeat of the Universal One. The 30 trillion pulsations of light per second exactly equals one super pulsation of our sun. The oscillation swing of the pendulum between the power of inertia and the equal power of motion is the fundamental principle of universal mechanics.

Energy of Thinking

The universal constant of thinking energy is behind all motion. It is available to humans in proportion to a person's desire. All energy is caused by universal thinking. All thinking is universal thinking, and all thinking is simultaneously universal. There is an apparent variability of intensity of thinking. This variability is due to the concentrative ability of mind to accumulate amounts of the universal thinking constant. Outer thinking (ego) is of the lower octaves and Inner Thinking (Higher Self/Inspiration) is of higher speed in the higher octaves. The greater the concentration, the greater is the power to create.

All things can be what they desire to be. Desire and effect evolve into the reality of that desire. The dimension of any wave of energy defines the dimensions of desire. Inertia cannot be overcome nor can motion take place without the concentrative desire force of the creative, imaging faculty of the mind.

Every particle of matter in the universe is connected with every other particle (entanglement) by electric charge, which controls opposing electromagnetic cones of energy. The apices of these opposing cones meet at the gravitational center of every particle of matter. The gravitational center of every particle of matter is the telegraphic center of communications, and the electric poles are the channels of communication. By means of this connection, every particle of matter in the universe is gravitationally and radiatively informed of any change of dimension in any other mass. All mass simultaneously responds to these communications by sympathetic contractive or expansive readjustment of its dimension.

All energy is created by thought. All energy expresses itself in motion. All motion is expressed in waves. All waves are sequentially evolved by electric action into accumulation of mass around minute whirlpool centers in the Universal Mind Substance. All whirlpool storm centers are formed by resistant magnetic reactions. All wave motions are effects. Dimension of motion within the wave is measured by the opposition of the electric pole and magnetic base.

To know the wave is to know the secret of creation. The wave is nature's method of transferring the dimensionless concept of form in inertia to dimensional form as mass in motion. All waves are the expression of a thinking force – a record of the universal constant. Waves are actions and reactions. Waves are simultaneously generated from inertia at intervals which are sequentially produced and reproduced in every atom of every element until they are recorded in the inertial planes. All of the energy of the universal constant or octave is existent in inertia and all its variations in accumulation are stored in the sequential inertial planes of equilibrium. The purpose of the wave is to transfer states of motion caused by thinking from one potential to another. **The purpose of motion is to contract concept into form and then to expand into memory of form.**

All concepts are but recollection of eternal ideas. All forms are but regenerated memories of Universal Mind. Every action of motion is recorded by a reaction in the master tone of each octave. The inert gases are zero in the universal equilibrium scale. Russell refers to inert gases as the spiritual element which birth and rebirth the physical elements. Each octave of an element grows from its inert gas as a tree grows from a seed. The inert gases record and store for repetition all that has gone before in that octave. Each action of motion is recorded by a reaction in the master-tone inert gas of each octave. They record, remember, and repeat all actions and reactions of everything for eternity.

The Illusion of Time

Time is the measure of universal repetitiveness. All phenomena of nature is repetitive. Without events, there would be no time. Intervals of time are an illusion, which gives birth to other illusions known as relations. Without intervals of time, relations would not exist. All dimensions are relative. Time is a dimension. Therefore time is relative. Time dimensions mark the periods of all phenomena of motion. Time belongs to motion and is not an attribute of Universal Mind Substance. In inertia of the Universal Mind Substance, time, like any other dimension, does not exist.

In the highest perceptible octave, the speed of reproduction of the universal constant is 186,330 miles per second. This is the speed at which energy overcomes inertia and sets it in motion in the highest octave of the lowest potential of matter perceptible on this planet. Each succeeding octave has its own time limit for reproducing the image of concept. The speed of light of the fourth octave is the expression of desire of electricity for action. Light appears and this illusion of motion gives us the universal standard we know as the speed of light. Time is but the recorder of change. It is the pendulum of motion. Time is one of many illusions that deceive man into believing that unreality is reality.

When an event takes place anywhere in the universe, the concept of that event is simultaneously measured in inertia through the Universal Mind of nondimension. This is the first step toward the production and reproduction of an event. The dimensionless universe of inertia in the Mind Matrix, the apparent nothing, has taken on the dimension of the apparent something. The illusion of space, time, and motion has appeared. Duration has also appeared. The thinking mind does create our physical world, which is but an illusion.

CREATION OF FORM

The Evolution and Dissolution Cycle

EVOLUTION

Thought

↓

Electro/Magnetic Energy (More Electric)

↓

Creates Various States of Motion in Mind Matrix

↓

Concentration and Desire

↓

Idea created

↓

Light Waves Become Particles

↓

Events and Form created

↓

DISSOLUTION

Loss of Concentration

↓

Form Destroyed (More Magnetic)

↓

Return to Universal Mind Matrix

↓

Memory of Idea in Universal Mind Matrix

↓

Stored as Knowledge

↓

The Cycle then Repeats Itself

↓

Back to Form

THE UNIVERSAL MIND MATRIX

Connected to Entire Universe

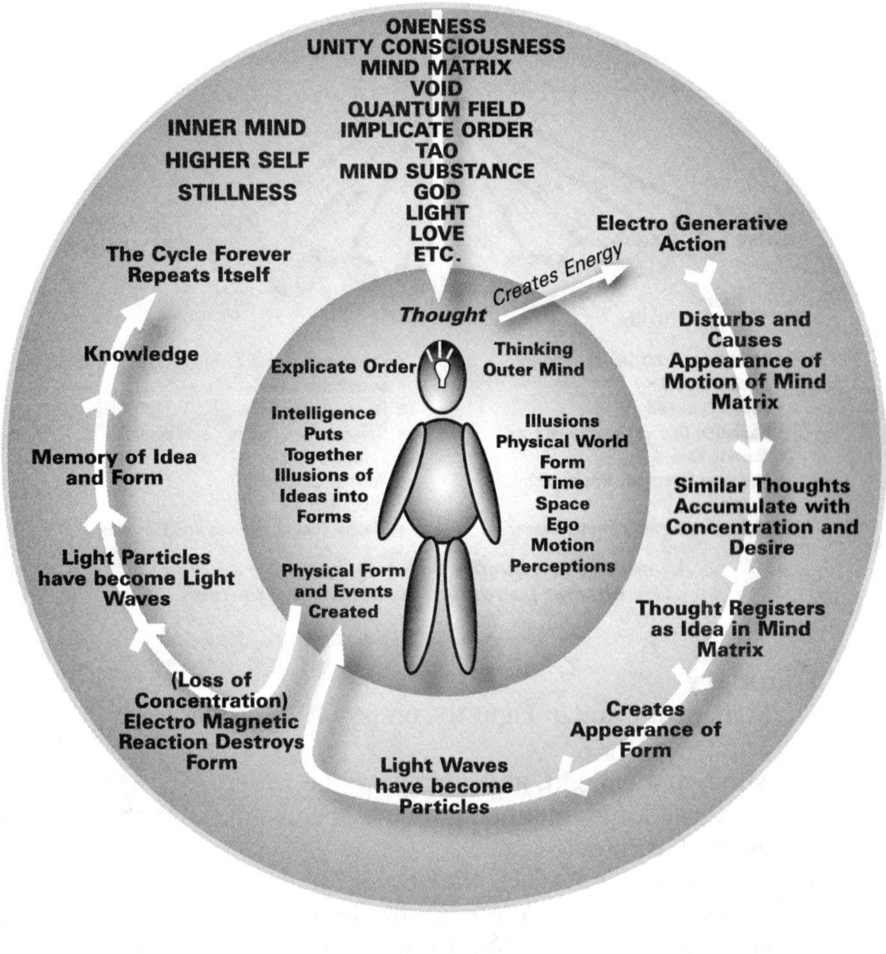

THE COSMIC CLOCK PENDULUM

Initiated by Thinking
10 OCTAVES

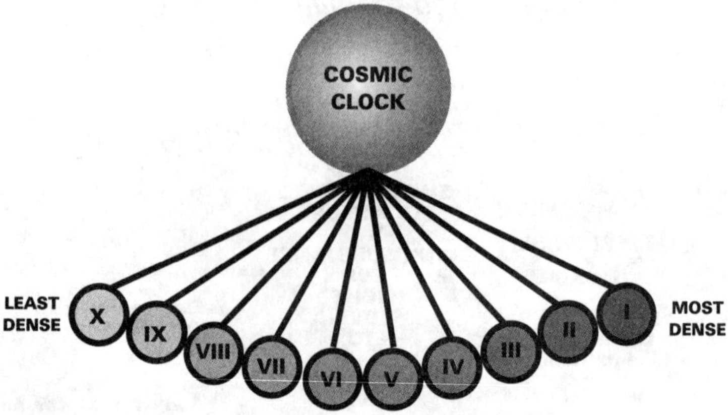

Octaves I to X
- **Accelerated Time**
- **Power Released**
- **Radiation Due to Centrifugal Force**
- **Form is Voided**
- **Negative Electricity (Magnetism)**
- **Exhalation**
- **Makes Light Particles into Waves**

Octaves X to I
- **Decelerated Time**
- **Power Accumulates**
- **Compression Due to Centripetal Force**
- **Form created**
- **Positive Electricity**
- **Inhalation**
- **Makes Thought Waves into Particles**

All energy is equalized in two opposite swings.
The cosmic pendulum forever repeats itself.

Summary

1. Man must think in Light to comprehend the Universal Mind.
2. Mind, matter, light, and energy are the same.
3. Universal Mind, matter, light, and energy are externally existent as cause.
4. The creating universe is the idea of Mind.
5. Mind thinks the idea into an illusion of form.
6. Form results from motion caused by thinking.
7. All motion is caused by thinking.
8. All motion is measurable, and therefore an illusion.
9. Illusions of form are varying states of motion.

10. All states of motion have ten wave octaves, and seven tones in each octave.
11. All creation can be reduced to 140 elements.
12. Motion is necessary to produce the form of an idea.
13. There are 18 dimensions.
14. The spectrum of the elements has been sorted into 7,000 colors of light.
15. The 7,000 lines of light are the letters of the universal language of Light.
16. When man thinks in Light, he knows he is One with God.
17. Man is omnipotent, omniscient, and omnipresent.
18. The thinking mind evolves ideas into appearances of form.
19. Thinking produces a force of electromagnetic light energy.
20. Positive electricity created by thinking produces gravity by compression.
21. Negative electricity produced by thinking produces radiation which degenerates.
22. All energy of the universe is caused by thought which disturbs the inertia of Universal Mind.
23. Thinking initiates the cosmic clock and cosmic pendulum.
24. The cosmic pendulum ticks off varying dimensions of all motion forever.
25. All dimensions of matter and energy are light corpuscles of Universal Mind.
26. Spirit is of high octave and matter of low octave.
27. All matter is crystalline, the apparent solidity of light.
28. Creation is a journey of the thinking mind into the illusion of the divine idea.
29. The sole purpose of mind is thinking, and ideas are the sole product.
30. The thinking mind evolves ideas into appearances of form.
31. All effects of thinking are omnipresent in the universe.
32. Thinking creates light corpuscles that register in the Universal Mind as ideas.
33. Light corpuscles are living beings that are both male (generative) and female (radiative), but one predominates.
34. All mass is both electric and magnetic.
35. Light corpuscles register the idea of thinking.

36. Light corpuscles assimilate ideas of other light corpuscles that evolve the idea.
37. Electricity is the attractive force of mind that compresses and accumulates mass.
38. Magnetism repels and causes dissolution.
39. The apparent opposites of electricity and magnetism are the cause of creation.
40. Electricity gives form to the idea and magnetism reduces form back to idea.
41. An idea is the eternal record of form.
42. Thinking is electromagnetism registered in Light.
43. Atoms differ from the solar and star systems only in dimension.
44. Each octave is a separate and complete sequence of inhalation and exhalation.
45. Inhalation of the Universal One is the cause of electricity and principle of growth.
46. Exhalation of the Universal One is the cause of magnetism and dissolution.
47. Every particle of matter in the universe is connected with every other particle by electric charge.
48. All energy expressed in motion is created by thought.
49. Waves of light are actions and reactions.
50. The purpose of waves is to transfer states of motion caused by thinking from one potential to another.
51. The purpose of motion is to contract concept into form and then to expand into memory of form.
52. All phenomena of nature is repetitive.
53. Without events, there would be no time.
54. Time is a dimension and relative.
55. When an event occurs, the concept is simultaneously measured in the nondimension of Universal Mind.

Chapter Twelve

THE LAW OF ONE

Man is One with God

God and Mind

The writings of Walter Russell help explain the concept of God and Oneness. The final two chapters reiterate and expand these concepts and how they relate to man. We have alluded to the reader that God, Universal Mind, Mind Matrix, quantum field, implicate order, Light, and Spirit are the same concept. If man is One with God, he also has attributes of omnipresence, omnipotence, and omniscience.

Russell said the greatest law of the universe is the Law of One. It states that "Everything that is, is of something else that is. Nothing is of itself alone. All created things are united." This is the law of the entire substance of Divine Mind, the Mind Matrix. It means there are no two independent separate things in the entire universe. (Quantum physics calls it entanglement.) Individuality does not exist. It is only an appearance of separateness and divisibility. The concept of Oneness is very simple. It is the spiritual substance of Divine Mind.

The One Universal Mind is a living substance of the one living, thinking Being of which all things in the universe are a part. That includes you and me. Everyone and everything. **Light is what makes the one substance a living substance. Light is the principle of Mind. It is the creative force of the universe, and Light is all there is. It is the living God. God is manifested in Light. Light is the Holy Spirit, God, and nature.** All created things are dependent upon light which holds them together in the appearance of individual or separate things. Light particles are the substance of all things that are in turn dependent on Light waves, thereby making them interdependent. Russell wrote:

> "All is God.
> God is all.
> God is Light.
> Light is God."

One who realizes this great law of Oneness knows love. **Love is a realization of the Oneness of the universe.** Once man knows this Law of Oneness in his heart, he will have no limitations.

Love is the ecstacy of inspired thinking. God's thinking is ecstatic thinking. When man knows this Law of One so that it resonates within his soul with ecstacy, inspiration, and contemplation, he will know the meaning of love. Once this happens, man will have unlimited power and true thinking. Inspired thinking is thinking in tune with Divinity. True thinking of the Inner Mind (Higher Self) renders true thinking of the outer mind (ego). **The Inner Mind of man is tuned to higher speeds of ecstatic thinking. Only the Inner Mind of man can think in Light.** The outer mind of man cannot know Light. When man thinks in Light, he will be a universal genius. There can be no imperfect ideas. There is no evil or imperfection. The appearance of evil is merely potential out of place and suspended there for time. Evil is an action which must eventually meet its reaction.

What man calls the "created universe" is the idea of thinking Mind. The ideas from Mind are perfect as Mind itself is perfect. Imperfection of ideas is impossible. With this knowledge, man can have dominion over all nature and all things. He can have dominion over his own body and command it to health and beauty. All knowledge will come to man when he knows the Law of One in his heart. All knowledge is existent in the universe and is the inheritance of thinking man.

Omnipresence, Omnipotence, and Omniscience

Man has been taught and trained to look for God outside

the realms of nature – a god that is substanceless, separate, and apart from himself. As we shall discover, God is within us all and in all phenomena of nature. **There is no other God than Mind. The only purpose of Mind is thinking. Ideas are the only product of thinking. Creation results from a thinking Mind.**
Russell wrote:

> "God is Mind.
> Man is Mind.
> God is man."

All ideas are registered in matter as form. Ideas are held in form by Mind for countless intervals of evolution and dissolution. All ideas of Mind are perfect. Imperfection of Mind cannot exist. Remember, Mind is God. The one Substance of Universal Mind is Light, and Light is universal. **Man is Light and therefore universal.** That which is universal is omnipresent. Therefore both man and matter are omnipresent. The Oneness of the universe allows no exception.

> "Mind is Light.
> Mind is all.
> One."

All of creation is contained in the desire of Universal Mind to express ideas and form. All Mind is universal, and, therefore, all Mind has all power. Omnipotence is defined as all the power that exists. **The mind of man is Universal Mind and therefore man is all power and therefore omnipotent.**

All this power is universally present throughout the entirety of this dimensionless universe. All matter and all creating things are nothing but images of thinking mind. That which is universal is not separate from itself. Because omnipotence is universal, all matter is omnipotent. All power lies in the thinking power of Mind. The whole universe is Mind which thinks in unison.

> "Mind is the universe.
> Mind is all.
> One."

All knowledge is within the thinking substance of Universal Mind. It can be said that the entire substance of Mind is Knowledge. Therefore knowledge is universal.

Man is in error when he thinks that the boundaries of knowledge and intelligence are beyond his comprehension. Knowledge and Intelligence of Universal Mind are limited and also simple. Knowledge is nothing more than the perception of dimensionless existence. It is both passive and inert. It belongs to the equilibrium of unchanging causes in the Universal Mind. The perception of knowledge is suspended within motion-in-inertia of the Universal Mind. The attribute of Universal Mind is Knowledge which is its reality. The alphabet can be used as a metaphor for knowledge. Within the alphabet is all knowledge, but until it is organized into words and phrases, the alphabet is without idea. Knowledge alone cannot be added to or subtracted from. Omniscience does not mean unlimited knowledge but means all the knowledge there is. Complexity of ideas does not constitute new knowledge. Man has all knowledge within himself. Inspiration will unlock the door to knowledge.

Because man is Mind and Mind is God, man is therefore omnipresent, omnipotent, and omniscient.

Illusion

Intelligence can be defined as understanding the reality of existence. Intelligence is registered in the illusion of existence. However, intelligence does not belong to substance but to motion. It belongs to the appearance of existence and to the unreality of the illusion of existence. Using the alphabet metaphor, intelligence is organizing the alphabet of knowledge into phrases of infinite ideas. It is the act of thinking creatively. **Intelligence puts together the illusions of ideas into forms of those illusions by thinking creatively.** Intelligence is not mere possession of knowledge. The desire for knowledge is the cause of the dynamic activity of intelligence. Constant within the limitations of the knowledge of Mind is the thinking of Mind. There is all intelligence in all matter.

All knowledge of the illusion of existence is limited to the perception of causes that create the illusion. The causes do not change. They are existent, and knowledge is limited to that which is existent. However, the effects are nonexistent illusions. They appear to exist, but are only ideas in transit. All ideas are transitory because the thinking Mind is always in motion and therefore changing. Man can have knowledge of the cause, but cannot acquire knowledge from changing effects. Knowledge is as simple as cause is simple. The effect can be comprehended even though it is complex.

There are no unconditional facts of matter in a universe of motion, but only appearances of facts. **Facts of form cannot be proven in a world of time, space, and motion**. Matter can be described as form in motion. It is an ever-changing effect of an unchanging Substance. The effects of matter are manifested in form and only fleeting and ever-changing. Therefore, you can say the facts of form in matter are but fleeting effects.

Causes of the appearances of existence are the only unconditional facts. Things are not what they seem in a universe of motion, as change is the only permanent thing. **All form is an illusion, because it is in transit between appearances and disappearances.** Transition cannot be considered existence, but only an effect of motion upon Substance. This Substance alone is existence. This substance is Universal Mind Substance (the quantum field).

An analogy for an illusion can be a ripple (form) on the water (mind). Ripples have no existence in themselves without the substance of water (Universal Mind). Without the substance of water, ripples could not appear. Having appeared in the form of ripples, they will disappear into the substance of water. That which appears must disappear. However, it must reappear in accordance with the universal law that no state of motion ever ends. **This divine concept is eternally repetitive in the reproduction of idea into form and its dissolution and then back again into the memory of form.**

All the effects of the creating universe can be comprehended only when all the causes are known. Man is unable to conceive of anything beyond the effect of motion and its cause. **Because of eternal repetition, nothing is or will be which has not always been.** No phenomena of matter can be a new phenomenon of matter. Finite limitations apply to all phenomena of matter. Every complexity of matter is but an effect of light units with orderly variance of motion. The forms or images of ideas are limited to the limitations of the ten-octave range of thinking.

Thinking is the cause of all effects! Because thinking is limited in range, the effects of that cause are limited in range. Likewise, comprehension is limited to the range of possible effects.

Illusions or appearances or form man mistakenly identifies as existence or reality. They are reappearances of old illusions. This is the universal Law of Repetitiveness. The orderliness of created things (form) is due to invariability of cyclic periodicity in all effects of thinking.

The Soul

The soul of man is the memory of the evolving idea of man. The soul is the record of man's thinking. From the soul is the body of man born again. The soul of a deceased person is the record of a person while he "sleeps" for awhile. Sleep and wakefulness are essentially simultaneous. They are expressions of their opposition, but predominantly one or the other in sequence. Death is only a longer sleep than daily sleep.

Reincarnation is the regeneration of the body. The soul's chemistry of all ideas is registered in the master tones, known as inert gases. The soul is the matrix of the body just as the master tones are the matrix of the physical elements. The inert gases of an octave (there are ten octaves with seven tones in each octave) contain a complete and exact record of every effect of motion in its octave. **Man's soul contains a complete and exact record of every action and reaction of thinking man**. Some refer to this as the Akashic Record.

Creation

A man who has only knowledge has only a negative possession. However, a man who possesses a great imagination has a positive possession. Knowledge is within all men. Knowledge of existence is recalled from within as a man "learns" new things.

As a man learns new things, he is recollecting old experiences of his thinking. Man thinks ideas into form as does the Universal Mind. All thinking is universal thinking. All thinking things are thinking in unison, and all are creating what they are thinking. In other words, man is his own creator.

It is when man knows the language of the Universal One of which he is a part that he shall know the voice of the Universal One. The still small voice within man speaks in the language of Light, in words of tones that travel at the speed of light. God created us so that we should fulfill His purpose. Desire was given to us to create as well as the power to create.

Memory

The acquisition of new knowledge by a person is but a recollection of old knowledge. Learning is recollection. Recollection is the act of transforming the potential energy of an idea into the kinetic energy of an active mind through concentrative thinking. Recollection can be described as the resurrection of the soul of an idea into the body of it. Essentially, memory is the storehouse of the idea of mind. The thinking mind taps into the storehouse of memory in the activity of mind that is called "memory," "instinct," "intuition," and "imagination." They are all the same; they are one.

Thinking of all mind is the same, whether it be the Mind of God, of man, of animals, etc. Mind is thinking in the expression of one idea, which is the whole, creating, and evolving universe. **God, man, the apple tree, and an atom are all one. All these are thinking out the divine One Idea.** Illusion of the One Idea is but an illusion in form, whose substance is the Universal Mind. Man too often believes illusion to be a reality. Man regards himself as one of many objects in the universe that is independent and is free to think as he desires. **When man understands the Law of One, he will realize that he is not separate from the plants, animals, and God.**

The Universal Mind also explains instinct. Animals have been making reappearances for ages, and all the acts of their thinking mind have been recorded as memory. This explains why animals know instinctively to migrate or build nests.

Imagination and Inspiration

The quality of imagination depends on the ability to recollect memories recorded in the soul, and the ability to think in Light. **To think in Light is to think in the higher octaves of the subjective Inner Mind (Higher Self).** To think in Light, one has to disassociate the objective outer mind (ego) from the Inner Mind. The outer mind is focused on continuing the body as an appearance of existence or survival. Man differs from animals only in his ability to think in higher octaves.

A great imagination reflects the great perception of the reality of universal existence and greater intelligence. Those who are inspired geniuses have great imagination with great intelligence. They are able to use knowledge creatively.

When man gives to others the inspiration which has come to him, he is a divine messenger of the living God. He delivers to mankind his revelation from the Holy of Holies. The prophets were great thinking men who were geniuses. They had knowledge of things and could foretell their effects. Genius is in the soul of everyone of us. Those who unlock it know its ecstacy. **No work of man can endure which does not originate from inspiration that was created by ecstacy.**

Those who attain the ecstacy of genius are the ordained messengers of the Universal One. They translate the word of the Universal One into the word of man for the soul of man. Genius, which is the soul, gives knowledge to man. A genius can make brutal men gentle. The message of genius is for the Inner Mind as the outer mind cannot comprehend it. Genius desires no reward and the glory of genius is humility, not arrogance. Walter Russell was such a genius.

In summary, man has all knowledge within himself. Inspiration is what unlocks the door to knowledge. The universe is omniscient and is of the electron, atom, man and the stars. Oneness of the universe allows no exceptions, no mistakes, and no errors. It is all inclusive. There is knowledge and intelligence in all matter.

"All matter is spirit.
Spirit is God.
God is universal.
God is all."

This is the Law of One.

Summary

1. All created things are united.
2. There are no two independent things in the universe.
3. Oneness is the spiritual substance of Divine Mind.
4. Light is the creative force of the universe. Thinking produces light.
5. God is manifested in light.
6. Light is the Holy Spirit, God, and nature.
7. Light is the substance of all things.
8. Love is the realization of the oneness of the universe.
9. Love is the ecstacy of inspired thinking.
10. Only the Inner Mind (Higher Self) of man can think in Light.
11. There is no evil or imperfection.
12. Ideas from the mind are perfect.
13. All knowledge will come to man when he knows the Law of One.
14. There is no other God than Mind.
15. The only purpose of mind is thinking.
16. Man is Mind. God is Mind.
17. All ideas are registered in matter and form.
18. Man and matter are omnipresent.
19. Man is omnipotent.
20. Mind is all.
21. All knowledge is within the thinking substance of Universal Mind.
22. The entire substance of Mind is Knowledge.
23. Man is Mind and therefore omniscient.
24. Intelligence is understanding the reality of existence.

25. Intelligence does not belong to substance but to motion.
26. Intelligence is the act of thinking creatively.
27. Intelligence puts together the illusions of ideas into forms of those illusions.
28. Matter is form in motion.
29. All form is an illusion because it is in transit between appearance and disappearance.
30. Because of eternal repetition, nothing is or will be which has not always been.
31. Thinking is the cause of all effects.
32. Man has control over his action but not to reaction over his actions.
33. Evil is an action that must eventually meet its reaction.
34. The soul of man is the record of man's thinking.
35. The soul is the matrix of the body.
36. Acquisition of new knowledge is but a recollection of old knowledge.
37. Inspiration unlocks the door to knowledge.
38. There is knowledge and intelligence in all matter.

Chapter Thirteen

ONENESS
What it Means

Evidence has been presented throughout this book that everyone and everything is connected by the Mind Matrix. If people understood this principle of Oneness and that thoughts create our perception of reality, perhaps they would take more responsibility for their thoughts and treatment of others. As mentioned earlier, the only real thing is our thinking mind. Everything else is impermanent and illusory. God is Mind and every one of us is part of that God Mind. We are all co-creators with everyone else in the universe resulting in an expanding universe.

Oneness means nonseparation. It means love. When we incarnated into this third-dimensional world, our soul knew that on a conscious level we entered knowing that we would be separated from knowing our Oneness with God. The purpose of life is to awaken this Oneness with God through our life experiences. What has kept us separated from the knowledge of oneness has been our egos. Although the ego has helped us survive in this world of separation, it is the ego that keeps us separated from Oneness. The ego has created the emotions of separation, such as judgement, fear, anger, criticism, greed, selfishness, sadness, guilt, and anxiety to name just a few. The book has provided the science underlying the concept of karma, cause and effect, the action and reaction of our thoughts and deeds. What we sow, we shall reap. All our actions will have an equal and opposite reaction. By understanding the Law of Oneness and The Law of Action (Cause) and Reaction (Effect), we have control of creating the illusion of separation when we treat others with love.

What is love? Webster Dictionary defines love as "an unselfish, loyal, and benevolent concern for the good of another." In other words," A brotherly concern for others," states Webster. Love is nonseparation. It is oneness. Love gives us peace of mind, our peace of God. If we understand the principle of Oneness, we will treat others as if they are one with us, which they are.

There is much wisdom in *A Course In Miracles* that Jesus dictated to Dr. Helen Schucman of Columbia University. Jesus said, "Love is All there is." This book has provided evidence that Mind is "All there is." Walter Russell wrote that God and Mind are One. He also said that "Love is All there is." God, Mind, and Love are but One. That the Inner Mind (Higher Self) thinks in Light, Oneness, and the outer mind (ego) thinks in separation. The Higher Mind is our Christ consciousness that is one with God.

A Course in Miracles states that the only real thing is love, and love cannot be threatened. It also states that nothing unreal exists, meaning the illusions of the physical world that our mind has created. The *Course* goes on to say, "the meaning of love is beyond what can be taught" and that "the purpose of the *Course* is to remove the blocks to love's presence, which is our natural inheritance." It also states that "the opposite of love is fear, but what is all-encompassing can have no opposite." Fear is but an illusion, an example of separation.

A *Course In Miracles* tells us how to remove these blocks of love, the separations. It is done by changing our perception toward events and things. **A change in perception is the miracle.** We all have a freedom of choice on this physical plane. We can choose to act with Oneness or separation. If an upsetting situation arises, we have a choice of how to respond. A choice of separation could be anger or fear. A choice of Oneness might be forgiveness or understanding. The choice is always available. A choice of Oneness will bring us peace. The *Course* also says that if you don't know how to respond to a situation, then you should ask the Father how to respond, and you will be directed.

This chapter will illustrate examples of Oneness that help eliminate the blocks to love. It will also suggest methods to achieve Oneness.

Examples of Oneness

Social Programs

We live in a world of competition and greed. Economists are telling us that the separation of wealth between the richest and poorest people is the greatest it has ever been. It is estimated that approximately 10% of the population owns 85% of the wealth. Does extreme wealth make a person happier? Social scientists say no. Most people on this planet struggle just to provide food and shelter to sustain themselves. These people have very little time for personal growth and education because survival is their main focus.

During the 20th century, humanity in developed nations began to realize that basic needs of people could be met by sharing wealth. Many of the indigenous people, such as Native Americans, had long ago discovered the principle of sharing. In Europe, many countries introduced social programs to their government structure to care for the less fortunate. They also believed proper health care was a right for everyone. In the United States, social welfare programs helped the poor, and Medicare and Medicaid provide health care to seniors and children in need. Social programs and public education are good examples of Oneness in society. It has been said that the level of evolvement of a society can be measured by how it cares for the less fortunate.

Charity

In recent years there have been numerous natural disasters around the planet. The outpouring of donations and supplies from around the world to these devastated areas has been remarkable. People in Haiti, Indonesia, Pakistan, and Japan have been appreciative of the generosity of people throughout world. It has shown that people are unselfish when it comes to helping their fellow man in need. These are acts of Oneness.

Americans are known for their generous contributions to nonprofit organizations to make life better for the less fortunate. Organizations such as the Red Cross, Salvation Army, United Way, World Hunger projects, Habitat for Humanity, and countless other charities depend upon the goodness of others. Charities are another example of how we show our nonseparation with others.

Volunteers

Volunteering is another good example of Oneness, as volunteers give their time and love. Hospitals around the world depend on thousands of volunteers to help provide the best possible service. Doctors without Borders is a good example of how health professionals volunteer their service to help humanity. Habitat for Humanity builds homes for the needy through volunteer help. One of the most successful programs of volunteering in the United States is the Peace Corps. As mentioned in the Tribute, Dr. Maury Albertson was the co-founder of the Peace Corps which has been in existence for fifty years. Young adults postpone their careers for two years in order to help citizens in third world countries better themselves.

The Environment

Indigenous people around the world have talked about their Oneness with nature. They knew they couldn't upset the balance of nature without harming themselves. Our industrial civilization has not understood this principle of Oneness with nature. As a result, pollution has contaminated the air, oceans, and land. It has destroyed forests, killed wildlife, and poisoned humanity. During the past fifty years, the consciousness of humanity has grown immensely regarding the environment. Once society realized that we need to live in balance with our environment, environmental laws were enacted around the world. Because of the carbon footprint of fossil fuels, scientists realized we need to develop alternative fuels to save our planet. Industry has found that being a green company is actually better for the bottom line, but they still have a long way to go. Humanity has awakened to the reality that we are one with Mother Earth.

Animal Rights

If we are One with everything and everyone, we are also One

with animals. This is one area that society has not awakened to, and our animals suffer. Everyone knows the unconditional love of our family pets. Dr. David Hawkins has proved this with his consciousness research. He has shown that dogs and other animals actully rated higher in consciousness than humans because of this unconditional love they possess.

Most readers will not want to read this, but eating animals is destroying our society. Americans eat an estimated 220 – 275 pounds of meat per person annually. It is damaging our environment, destroying our health, evoking mental health problems, causing poverty, provoking wars, and causing undue mental stress to our animals. How can this be?

Raising livestock for food is very energy demanding. We use a significant percent of total energy consumption in raising food for our animals. If we did not use this energy for raising food for livestock, we would have less of an energy crisis. Air pollution would be reduced, and contaminated water from fertilizers and pesticides would be reduced. The livestock meat industry is the greatest cause of global warming. About 35 percent of the world's greenhouse gases is caused by livestock and the methane it produces. This is more than all the greenhouse gases produced by all forms of transportation. There are millions of people in the world who are starving, and this food that we feed to animals could help to eliminate human hunger.

Did you know that approximately 67% of all health problems are related to animal based diets? Science has shown animal based diets can cause cancer, heart disease, atherosclerosis, strokes, hypertension, and most major diseases at some level. Good health and longevity depends on eliminating an animal based diet. A ten year National Cancer Institute study found those who ate the most red meat increased their risk of death by 30 percent. Men who were big meat eaters increased cardiovascular disease by 27 percent and women who were big meat eaters increased cardiovascular disease by 50 percent.

A very insightful book about eating animals was written by Michio Kushi entitled *One Peaceful World.* Kushi analyzed the diets of various societies going back millennia. He then correlated the diets of these civilizations to their warlike tendencies. He found societies who ate meat were more warlike. This correlation between a meat-based diet and war has extended throughout all of history. For example, the Hopi Indians were a very peaceful tribe whose diet was basically plant based. The Blackfoot tribe depended on buffalo for their diet, and they were one of the most warlike tribes. Kushi believes that by eliminating diets based on animals, we would indeed have one peaceful world.

About 95% of meat in America comes from factory farms. Conditions for these animals is horrific. There is great suffering of factory farm animals during their short life. They spend their entire lives in confined pens separated from their mothers and often denied all socialization. Most poultry is housed in crowded cages where the birds can barely move. Many have been genetically modified to have large breast meat with grotesque bodies. Feedlot cattle are crowded together in pens with dirt and feces underfoot. They are injected with antibiotics and growth hormones. Some scientists speculate that growth hormones in animals is causing the obesity epidemic in America. The antibiotics given animals may be causing the resistence to antibiotics that so many people are experiencing. Being led to slaughter causes an enormous stress to animals. Dr. Temple Grandin of Colorado State University has designed slaughter houses so the animals do not have as much fear when being led to slaughter.

We are One with the animals. The fear and stress of factory-raised animals is registered in the universal Mind Matrix. Hunting animals also creates a great stress for the hunted and this also registers in the Mind Matrix. How many animals are wounded and suffer drawn-out death experiences? Fish suffer when hooked in the mouth and are dragged through the water and then suffocate once removed from their environment. This is the sport of fishing.

Animal rights groups are raising consciousness about the way we treat animals, but most people don't want to hear this because meat tastes so good. Meat eater Andy Rooney of "60 Minutes" admitted eating fellow creatures which is working on his conscious. He thinks that within 50 to 100 years civilized people will refuse to eat animals. Many people are becoming vegetarians as they hear about the abuse humans do to animals. We would become a more healthy and peaceful society if we gravitated away from an animal-based diet. I have often thought that perhaps it is a karmic price that society is paying for the way we treat animals, i.e, health problems, environmental problems, and peace problems.

World Conflicts

I am hopeful that humanity will be able to put an end to war as our consciousness increases. (Please read the epilogue) Europe discovered that after centuries of war, there must be a better way. The European Union was formed and what a great example of Oneness! Economists have discovered there are very few conflicts between nations who trade with one another. Perhaps a global economy is what we need to prevent wars.

Another good example of Oneness among countries is the United Nations. Although it is not perfect, the UN has prevented many wars and conflicts. Nations have a forum to present their concerns to help resolve disagreements before they lead to war.

With all the conflicts in Africa, is it not wise to think of an African Union? In fact this is happening, and hopefully may provide a solution to the African woes. Perhaps a South American Union could help the economies of South American countries. What would happen if we had a democratic World Union? If the World Union was of the Light, and not the dark that many conspiracy theorists are afraid of, we could have a world of peace. With all the world changes and globalization taking place, maybe a World Union might be the way to a millennia of world peace that the Bible has prophesied. This would be the ultimate in Oneness consciousness regarding our many diverse nations.

Religion

In my opinion, there are many paths to God. Each culture and society has a religion to help guide their people to Oneness with God. In most religions, the core message is the same. They are saying the same thing but in different words. All religions want world peace and the best life for their followers. However, historically, most wars have been fought over religious differences. It often has been the egos of religious leaders that has caused the wars, not the religious teachings. Most religions teach peace, tolerance, compassion, and love. In the epilogue, I propose a "Oneness Project" that uses the scientific principles in this book that will bring all the religions together to help create world peace.

Minorities

Some of the greatest strides in Oneness consciousness during the past 150 years have been the acceptance of all people as equals. In America, it began with the Civil War and the freeing of slaves. There have been vast improvements with acceptance of black people as equals by the white people. There still is a way to go but it is happening. The next major change in America was suffrage and the equality of women. This happened with voting privileges, and continues today with equal employment opportunities. Eventually equal pay will happen. Today in America, the percentage of women receiving college degrees is greater than that of men. Gay rights is today's challenge and more and more states are accepting gay marriages. No matter what race, sex, or sexual preference, we are all One people. This is the lesson that many people came here to learn.

The consciousness of humanity is evolving as demonstrated by Dr. Hawkins' research. More people are discovering that what is good for the whole of humanity is also good for the self. As more and more people understand and practice the concept of Oneness, our health, environment, happiness, and world peace will all benefit. Our civilization has made great strides during the past century in discovering the rewards of Oneness.

Achieving Oneness

Examples of Oneness have been presented suggesting that Oneness is viewing the world as one large interconnected family. By being of service to others without expecting anything in return, we are sharing our love with our one-world family. This is Oneness. This is nonseparation.

We know we are not in a state of Oneness when we have feelings of anger, fear, shame, guilt, hate, selfishness, revenge, and anything else that does not give us peace. We all have had negative feelings and sometimes find them difficult to release. Psychologists believe the separation experience can be reduced to one of three universal fears, which include abandonment, low self-worth, or a lack of trust.

How do we overcome these negative feelings of separateness? We need to change our perceptions. We have a choice about how we react to something. Our reaction produces another action that keeps the cosmic pendulum swinging. If it is a positive reaction from the Higher Self, it will probably bring us peace. If it is a negative reaction, the ego has influenced our decision that puts us in the victim mode. Feelings of anxiety, fear, depression, and anger may transpire. The choice of our reaction is between Oneness and separation. If one doesn't know how to respond in a situation, we can ask the "Source" for direction for the appropriate response.

One of the early pioneers who realized the power of the mind was Dr. Norman Vincent Peale, a spiritual leader in New York City who wrote the book entitled *The Power of Positive Thinking*. The book has helped countless people through difficult times by helping them change their perception of negative situations. Positive thinking changes the attitude of perception into alignment with Oneness because we have a choice of how we react to a negative situation. A positive attitude is seeing the good in everything and a chance for personal growth in a negative situation. It is treating people with love, not anger. Those who have decided to follow this path, including myself years ago, find that it takes a concerted effort to practice the principle of positive thinking. Eventually, one reacts positively to a negative situation and it becomes natural, like a habit. The tool that Dr. Peale advised people to use was prayer to help change negative attitudes into positive attitudes.

The desire of our thoughts is what creates our perceived reality. It is the act of focusing one's consciousness that drives an act of creation. Consciousness is what creates. Desire is the focusing of our consciousness. If we have a positive attitude and confidence about something that we want or an event we want to happen, it will happen. Numerous scientific studies have been conducted to prove the power of prayer, especially with healing. Prayers are thoughts, and thoughts are prayers. To be a co-creator, we need to focus our thoughts. Gregg Braden wrote, "The universally connected hologram of consciousness promises that the instant we create our good wishes and prayers, they are already received at their destination."

It does not take a large number of people to initiate a change in group consciousness. Transendental Mediation (TM) suggests that it takes less than the square root of one percent of the population to change group consciousness. It is the positive thoughts of Oneness by a small segment of the population that can change the consciousness of the whole, as shown in crime reduction in TM studies. We must practice the concept of non-separation and treat everyone with love if we want to experience a world in peace. We must also live the ideals in our lives of the very things that we choose to experience in our world.

Walter Russell wrote about the greatest secret in the world. The secret being that "Love is something which must go out from the Light and find expression. In going out from the Light and finding expression in nature everywhere, the lesson is that Love is not out there, but in oneself. That is all that exists. When one knows it and has it, it is the greatest secret in the world, the secret of all things, the secret of mastery of oneself, the secret of uniting oneself with every other self through the forms of Love, through the power of the Light that shines from the eyes of anyone who has been illumined, anyone who found his immortality and desires to bestow his immortality upon every other human."

Russell said that it is the great urge of all humankind to manifest that Love in material form. It is the desire force of the soul, which is all knowing, to express itself by thinking in Oneness within the material world.

Russell was asked, "Where shall I find the Light?" He said you can look forever where it is not. But you can look where it is and find it in a flash of a second. You will be transformed into a God-being who knows Love. When that happens, man will know his immortality. He said that it is the greatest desire of all immortals to bestow this immortality on other humans.

Russell was then asked, "Who are the immortals?" He said it was those who have learned to Love. It is those people who give out all the Love they have in them, according to the Law upon which the universe is founded. It is those who know Love outside of matter, outside of material things, outside of form, outside of human beings, and outside the created universe. "They know love is not in any of those things," said Russell. "It is deeper than all of those things." Those who know their immortality can bestow their immortality onto others to the extent which others are ready for it. He who understands Love understands the rhythm of Heaven, wrote Russell. The immortal opens the door and uplifts man to higher standards.

Those who have discovered their immortality love their work and life, and in loving it, they love all there is of themselves. By awakening in them the ideal of Oneness and Love, they want everyone to know that it is also within them. And Love is the secret of it all. We are all One.

Russell said that one who has known God within him has found the Kingdom of Heaven. He knows of his Oneness with the Light, and he knows of his universality with all mankind and a unity with all creation. When one knows the rhythm of Heaven in him to the exclusion of everything else, only then can man know God. Only then can man know Love.

Love, wrote Russell, is the basic principle from which the universe is created. Love is the basic principle of creation and of the creator. It is the only power in the world. He said Love, Knowledge, and Power are the same thing. If you have one of these, you will have all of them. He also wrote that Love and Law, Life and Truth, and Balance all mean the same thing. "When one has one, one has them all."

Russell wrote that no one can say, "God gave another more than he gave me." All one can say is "The other is more aware than I, and I desire that awareness." When one desires that awareness to any extent, it is his. If he desires it without any doubts, it is his. Every man is limitless in the attainment of power and can attain power to the extent to which he knows Love. He can do this because he understands The Law of One.

EPILOGUE

Proposal for the Oneness Project

This book has explained how powerful the mind is, suggesting that the thinking mind may be All That Is. Evidence has been given that when a relatively small number of people focus on something, such as reducing crime in a particular city, the crime rates drop significantly. An amazing experiment took place on December 31, 1986, that was organized by John Randolph Price. He coordinated a world-wide meditation for world peace to occur on that day at 12:00 P.M. Greenwich Time. It was estimated that more than several million people participated in the peace meditation that day. During the time that followed the peace meditation, the world saw the end of the Berlin Wall and the end of the Cold War with the collapse of the Soviet Union. During the following year newspaper headlines literally read "Peace is Breaking Out Around the World." Whether this was a coincidence can be debated, but many believe it was connected to the December 31 Peace Meditation.

We now have the knowledge and science that strongly suggests that humankind has the ability through group consciousness to make this a peaceful world. Nearly everyone in the world desires world peace.

How can we make this happen? One suggestion is that we ask the major religions to come together and dedicate one day to pray and meditate for world peace. The enormous number of participants and the great power of these unified minds could create a consciousness field that would almost guarantee world peace. The scientific hypothesis underlying this experiment is undeniable. The questions that need to be answered are:

1. Is it realistic to get major religions to cooperate on such a global scale?

2. How can we make this happen?

The first challenge is to recruit a world leader who would share the passion for such an event. The leader could be a figurehead who would lend his or her name and be respected by all religions and the world community. Some possibilities might be the Dali Lama, Nelson Mandela, Jimmy Carter, or President Barack Obama who are all dedicated to world peace and are Nobel Peace Prize laureates.

The next step is to establish a working committee, the "worker bees" to make this happen. Another suggestion is to establish a committee of religious leaders to form an alliance for world peace. Hopefully, each religion could provide a representative, such as the Pope or perhaps somebody like Desmond Tutu or the Dali Lama. Other world leaders that might be recruited would be past Nobel Peace Prize laureates. Another suggestion is to ask the United Nations to assist in this great experiment. If enough world leaders came on board, the news media around the world could generate enough publicity to make this truly a global event. The cost would not be great. Perhaps a philanthropist would step forward to seed such a project. The Oneness Project is an organization under the International Association for New Science, a 501-c-3 nonprofit corporation.

The potential for world peace is enormous, if we could get all the world's religions to cooperate. All religions want world peace, and this would be a time for them to step forward to practice this message. This event will happen if enough of us can visualize a World Prayer Day for Peace and for the right leadership to come forward.

If any of the readers have ideas or suggestions, please contact me at:

The Oneness Project
Robert Siblerud
9435 Olsen Court
Wellington, Colorado 80549
Rsiblerud@aol.com
(970) 568-7323
Thank You!

REFERENCES

Binder, Tim, *In the Wave Lies the Secret of Creation*, University of Science and Philosophy, Waynesboro, VA, 1995. Chapter 11.

Braden, Gregg, *The Divine Matrix: Bridging Time, Space, Miracles and Belief*, Hay House Inc., Carlsbad, CA, 2007. Chapters 1, 2, 3, 5, 13.

Briggs, John and Peat, David, *Looking Glass Universe: The Emerging of Wholeness*, Simon and Shuster, New York, 1984. Chapters 1, 3, 5.

Capra, Fritja, *The Tao of Physics*, Bantum Books, New York, 1975. Chapters 1, 4.

Clark, Glen, *The Man Who Tapped the Secrets of the Universe*, The University of Science and Philosophy, Swannanoa, VA, 1946. Chapter 10.

Dossey, Larry, *Recovering the Soul*, Bantum Books, New York, 1989. Chapter 13.

Fillmore, Charles, *Atom-Smashing Power of the Mind*, University School of Christianity, Unity Village, MO, 1949. Chapter 9.

Friedman, Norma, *Bridging Science and Spirit*, The Woodridge Group, Eugene, Oregon, 1993, Chapter 8.

Hawkins, David, *Power Vs. Force: The Hidden Determinants of Human Behavior*, Hay House, Carlsbad, CA, 1995. Chapters 7, 13.

Holmes, Ernest, *The Science of the Mind: A Philosophy, A Faith, and A Way of Life*, Tarcher/Putman, New York, 1938. Chapter 9.

Jahn, Robert and Dunne, Brenda, *Margins of Reality: The Role of Consciousness in the Physical World,* Harcourt Brace Jovannovich Publishers, San Diego, 1987. Chapter 5.

Kushi, Michio, *One Peaceful World,* St. Martin's Press, New York, 1987. Chapter 13.

McTaggert, Lynne, *The Field: The Quest for the Secret Force of the Universe,* Harper Perennial, New York, 2001. Chapters 1, 5.

Milanovich, Norma, *We the Arcturians: A True Experience,* Athena Publishing, Scottsdale, AZ. Chapter 8.

Peale, Norman Vincent, *The Power of Positive Thinking,* Ballantine Books, 1952. Chapter 13.

Russell, Walter, *The Universal One,* University of Science and Philosophy, Waynesboro, VA, 1926, 1974. Chapters 10, 11, 12.

Russell, Walter, *The Secret of Light,* The University of Science and Philosophy, Swannanoa, Waynesboro, VA, 1947. Chapter 10.

Russell, Walter, *The Immortality of Man,* The University of Science and Philosophy, Swannanoa, Waynesboro, VA, 1991. Chapter 13.

Schucman, Helen, *A Course in Miracles,* Foundation for Inner Peace, Tiburon, CA, 1975. Chapter 13.

Siblerud, Robert, *The Science of the Soul: Explaining the Spiritual Universe,* Sacred Science Publications, Wellington, CO, 2000. Chapter 8.

Swanson, Claude, *Life Force: The Scientific Basis,* Poseidia Press, Tucson, AZ, 2010. Chapters 5, 6, 7.

Talbot, Michael, *Beyond the Quantum,* Bantum Books, New York, 1986. Chapters 1, 2, 5, 8.

Talbot. Michael, *The Holographic Universe*, Harper Perennial, New York, 1991. Chapters 1, 2, 3, 5, 8.

Wilber, Ken. *The Holographic Paradigm and Other Paradoxes*, New Science Library, Boston, 1985. Chapters 3, 8.

Winter, Mary, *PETA May Have a Point*, Denver Post, May 15, 2011. Chapter 13.

INDEX

For Additional Information

Walter Russell
The University of Science and Philosophy
P.O. Box 520
Waynesboro, VA 22980

Claude Swanson, Ph.D.
Synchronizeduniverse.com

David Hawkins, M.D., Ph.D.
Veritas Publishing
P.O. Box 3516
Sedona, AZ 86340